TRANSFORMER EXAM CALCULATIONS
By Tom Henry

Copyright © 1989 by Tom Henry. All rights reserved. No part of this publication may be reproduced in any form or by any means: electronic, mechanical, photocopying, audio/video recording or otherwise, without prior written permission of the copyright holder.

While every precaution has been taken in the preparation of this book, the author and publisher assumes no responsibility for errors or ommissions. Neither is any liability assumed from the use of the information contained herein.

This book was written as a study-aid for an *electrician* preparing for an electrical examination. There has been a definite need for one single book on transformer calculations presented in a simplified, easy-to-study format for an *electrician* to understand.

A very troublesome part of the electrical examination for many. In preparing this study-aid book the thought that was kept foremost in my mind was to condense the explanation and to include as much material as possible that would be of practical value to the *electrician* for the examination.

As you read this study-aid book on transformer calculations you will note that complicated electrical formulas and explanations have been put in a clear, concise understandable language for the *electrician*.

"Written for an electrician by an electrician".

Tom Henry

DISCARDED

NOV 4 2024

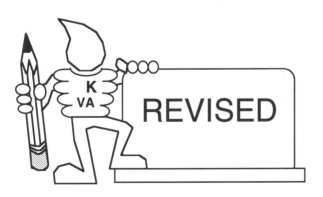

ISBN 0 - 945495 - 14 - 5

Asheville-Buncombe
Technical Community College
Learning Resources Center
340 Victoria Rd.
Asheville, NC 28801

CONTENTS

A transformer is a basic and very useful device in AC circuits.

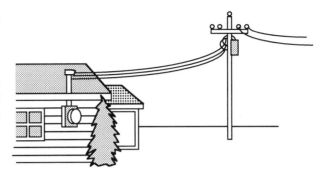

The first transformer built in America was in the year 1885 over 100 years ago. Now, so widely used you would rarely have to walk more than 200 feet to find one.

A transformer *does not* generate electrical power, it *transfers* electrical power. A transformer is a voltage changer.

A device, usually consisting of two insulated windings on a common iron core, in which alternating current is supplied to one winding and by electromagnetic induction which induces alternating EMF'S in the other winding.

One of the windings is designated as the *primary* and the other winding as the *secondary*. The primary winding *receives* the energy and is called the *input*. The secondary winding *discharges* the energy and is called the *output*.

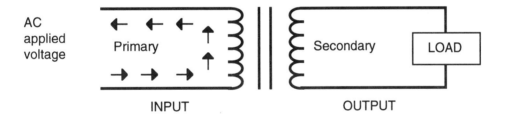

The transformer works on the principle that energy can be efficiently transferred by magnetic induction from one winding to another winding by a varying magnetic flux produced by *alternating current*.

An electrical voltage can only be *induced* while there is a relative *motion* between a wire, or a circuit, and a magnetic field. *Alternating current* provides the motion required by *changing direction 60 times a second.*

Direct current (DC) is not transformed, as DC does not vary in its amount from one second to the next. This is the reason we wire buildings with AC, we can't transform DC.

1

The two windings, primary and secondary are *linked* together with a magnetic circuit which must be common to both windings. The *link* connecting the two windings in a magnetic circuit is the *iron core* of which both windings are wound. Iron is an extremely good conductor for *magnetic* fields. The core is not a solid bar of steel, but is constructed of many layers of thin steel called *laminations*.

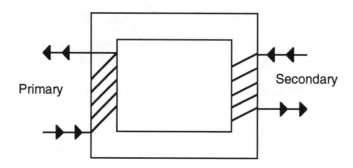

Since the primary and secondary are wound on the same iron core, when the primary winding is energized by an AC source, an alternating magnetic field called *"flux"* is established in the transformer core. The flux surrounds both the primary and secondary windings and *induces* a voltage in them.

Since the same flux cuts both the primary and secondary windings, the same voltage is induced in *each turn* of each winding.

The voltage in the primary will be the same as the supply voltage, but the secondary voltage will depend on the number of *turns* in the secondary winding in proportion to the number of *turns* in the primary winding.

If the secondary has the same number of turns as the primary, the voltage will be the same in both windings.

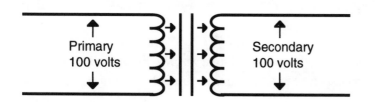

1 to 1 ratio
Primary = 100 turns = 100 volts
Secondary = 100 turns = 100 volts

By changing the *ratio* of turns in windings you change the voltage.

If the secondary has *half* the number of turns as the primary, the secondary voltage will be *half* as high as the primary voltage.

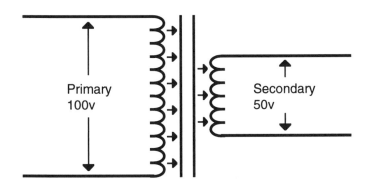

Voltage is stepped down
at a **2 to 1 ratio**

Primary = 100 turns = 100 volts
Secondary = 50 turns = 50 volts

If the secondary has *twice* the number of turns as the primary, the secondary voltage will be *twice* as high as the primary voltage.

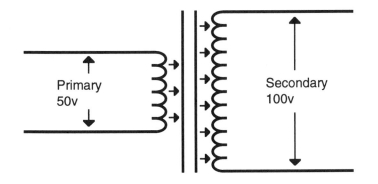

Voltage is stepped up
at a **1 to 2 ratio**

Primary = 50 turns = 50 volts
Secondary = 100 turns = 100 volts

The voltage is *stepped down* when the primary has more turns than the secondary. The voltage is *stepped up* when the secondary has more turns than the primary.

When the primary winding and the secondary winding have the *same* number of turns, there is no change in voltage, the ratio is 1/1 or *unity*.

The ratio between the voltage and the number of turns on the primary and secondary windings is called the *turns ratio*.

Np = number of primary turns
Ns = number of secondary turns
Turns Ratio = Np/Ns

Ep = primary voltage
Es = secondary voltage
Turns Ratio = Ep/Es

It is customary to specify the ratio of transformation by writing the primary (input) number first. Example: 30 to 1 is a step-down transformer, whereas a 1 to 30 would be a step-up transformer.

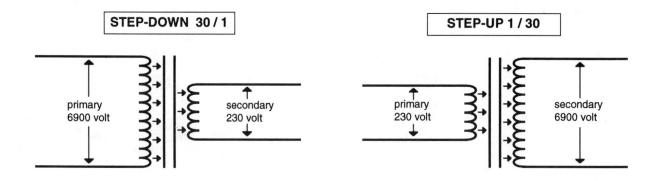

STEP-DOWN 30 / 1

primary
6900 volt

secondary
230 volt

STEP-UP 1 / 30

primary
230 volt

secondary
6900 volt

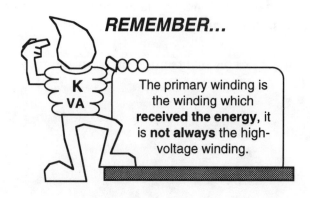

REMEMBER...

K
VA

The primary winding is the winding which **received the energy**, it is **not always** the high-voltage winding.

The purpose of changing voltages is that conductors are sized to amperage, by raising the voltage the amperage is lowered, thus the wire size is smaller.

A dynamo or generator at the main power plant is limited on the output voltage due to commutation with brushes, slip rings, moving parts, etc. Flashovers at high-voltage can cause serious damage. By using a step-up transformer we can take the output voltage of a generator and step it up to extremely high transmission voltages to 200-500 kv to supply power to cities.

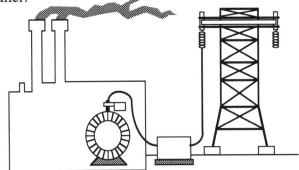

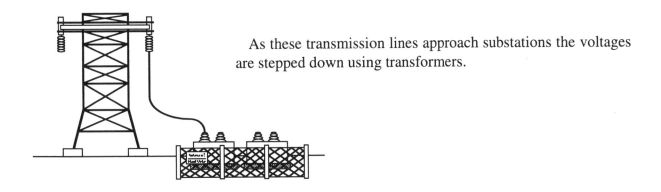

As these transmission lines approach substations the voltages are stepped down using transformers.

As the high-voltage approaches buildings it is stepped down again with a transformer to safe level voltages that you utilize in your home.

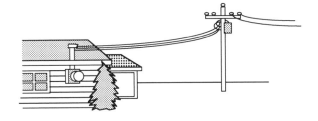

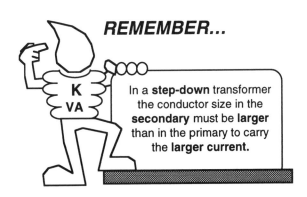

REMEMBER...

In a **step-down** transformer the conductor size in the **secondary** must be **larger** than in the primary to carry the **larger current.**

The transformer is our most efficient piece of electrical equipment, by having *no moving parts* there is no friction. A heavily loaded power transformer has been known to be 99% efficient. The losses in a transformer are copper losses in both windings and eddy-current and hysteresis losses in the iron core from magnetizing and de-magnetizing 60 times a second.

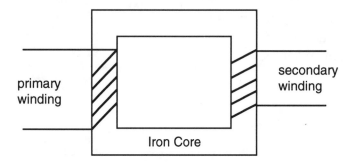

The magnetic flux may link the windings either through an iron core or an air core, the air core being used for high frequencies as with radio frequencies.

Eddy currents are caused from the flux changing in the iron, this generates a voltage in the laminated paths which set up a flow of currents at right angles to the direction of the flux path. Hysteresis loss is the energy required to reverse the direction of the flux in the magnetic circuit. These two components make up the iron loss.

Copper loss is the resistance loss or *heat* loss, is a result of the resistance in the transformer winding. Copper loss and iron core loss must be combined to find the total loss of the transformer.

The efficiency of a transformer is the ratio of the output to the input.

EFFICIENCY = OUTPUT/INPUT   OUTPUT = SECONDARY   INPUT = PRIMARY

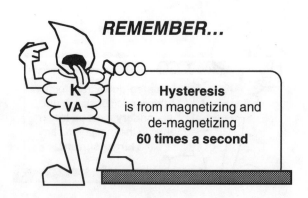

REMEMBER...

Hysteresis
is from magnetizing and
de-magnetizing
60 times a second

Often referred to as the *isolation type* transformer, as there is no mechanical connection between the windings, the primary and secondary windings are isolated from each other, being connected only by a magnetic flux.

When a voltage *less* than the rated voltage is applied to the primary winding, the flux is below design level. Induced voltages tend to follow but will decrease sharply when the transformer supplies current to the load.

If voltages *higher* than the rated voltage are applied to the primary, the excitation current increases, the magnetic core saturates and the induced voltage cannot increase. This causes a greater difference between induced and applied voltage in the primary, causing an increase in primary current that overheats the transformer.

For a magnetic field to produce a given number of magnetic lines, a current of a certain number of amperes must exist in the circular loop. This unit is called the *ampere-turns* and is defined as the unit of magnetomotive force which is found by multiplying the current in amperes by the number of turns in a winding.

If a transformer is operated above its power rating the efficiency will be poor and the transformer will run hot. Insulation can burn causing short-circuits between the winding turns.

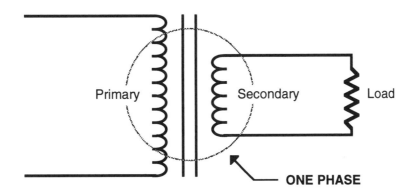

With the isolation type transformer the load is connected to the secondary winding in step-down transformation.

AUTOTRANSFORMER

The isolation transformer has separate primary and secondary windings which are electrically insulated from each other. The *autotransformer* has a primary and secondary winding in the form of one continuous winding on the *same* core. Therefore, the primary and secondary sections of the winding are in the same magnetic circuit.

The autotransformer is a *single* winding transformer.

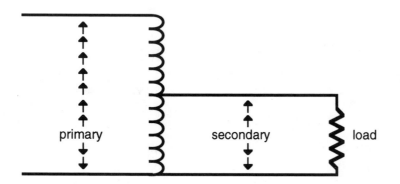

When voltage is stepped **down**, all of the winding is the primary, and part of the winding is the secondary.

When voltage is stepped **up**, part of the winding is the primary, and all of the winding is the secondary.

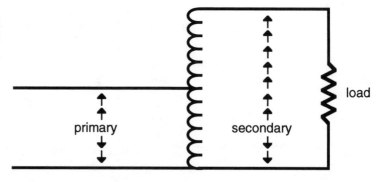

The principal use of an autotransformer is in motor starters where they apply half the supply voltage to the motor to reduce the current surge on starting. This is called *autotransformer type starting*.

Although autotransformers are used most exclusively for step-down applications, occasionally they are used as step-up transformers in lighting fixture ballasts and to boost the voltage from 208 volts to 230 volts.

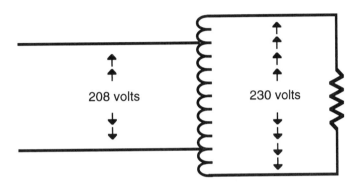

When the *ratio of transformation is small* the autotransformer is the *most efficient* and effective. *Less copper* is required for an autotransformer than for a two-winding transformer.

Not economically desirable when the ratio of transformation is *greater* than 2 to 1. Another disadvantage is that the high-voltage and low-voltage systems are electrically connected and could become hazardous to equipment and personnel. The Code forbids the use of autotransformers for lighting and appliance branch circuits unless there is a grounded conductor common to both the primary and secondary circuits.

The interconnection of the primary and secondary windings permits some of the power required by the load to be conducted from the supply to the load. This is contrast to the standard two-winding isolation transformer, it must transform all the power of the load.

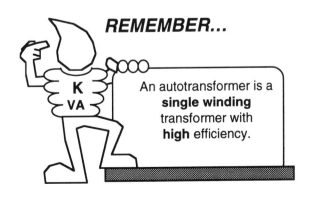

REMEMBER...

An autotransformer is a **single winding** transformer with **high** efficiency.

FOUR CLASSES OF TRANSFORMERS

• CONSTANT-POTENTIAL TRANSFORMER

A constant-potential transformer is used for the transmission and distribution of power.

• VARYING-POTENTIAL TRANSFORMER

Used for gaseous-discharge lamps where it is necessary to reduce the secondary with an increase of load.

• CURRENT TRANSFORMER

The current transformer is designed to change the current of a system. Used for instrument transformers or street lighting.

• CONSTANT-CURRENT TRANSFORMER

Designed to supply a constant current to the secondary circuit. The voltage of the secondary will increase or decrease according to the load but the secondary current will remain constant through a moving coil regulator. This is used in a series street lighting system.

INSTRUMENT TRANSFORMER

Used to step voltage or current down so that it can be used safely for instruments, such as voltmeters, ammeters, wattmeters, etc.

An instrument transformer has a small va capacity.

Accuracy is the upmost requirement of voltage and current transformation when connected with the metering of power.

Instrument transformers measure high currents and high voltages with low-scale ammeters and voltmeters.

There are two basic types of instrument transformers, the *current transformer* and the *voltage potential transformer.*

CURRENT TRANSFORMER (CT) (Doughnut)

Also called the series transformer because the primary usually has one or more turns connected in series with the line.

A current transformer is used when AC current is so large that connecting measuring instruments such as a kwh meter would be impracticable. The CT provides a means of reproducing the effect of the primary current on a *reduced scale* suited to the kwh meter.

Current transformers are used with power-factor meters, watthour meters, trip coils of circuit breakers, AC ammeters, etc.

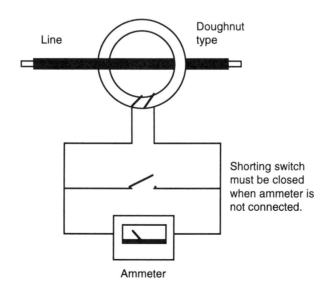

11

To standardize current devices the *secondary* of a current transformer is always *rated* at 5 amperes, no matter what the ampere rating of the primary is.

The current *rating* of the primary is determined by the maximum load current to be measured.

Example: Maximum load current to be measured is 500 amperes. The secondary winding will have a *rating* of 5 amperes. The ratio between the primary winding and the secondary winding is 5/500 = 1 to 100.

Thus, the secondary winding will have 100 times as many turns as the primary.

Using this 1/100 ratio, a current transformer for a load of 400 amps, the secondary would read 4 amps. For a load current of 300 amps the secondary would read 3 amps.

The ammeter is reading a *percentage* of the load current on a *low-scale* ammeter due to a current transformer (CT).

Some current transformers have only a solid bar (one turn) as a primary connected in series with the load to be measured.

With the doughnut type (CT), the conductor passing through the center magnetizes the core because of magnetic lines around the single (one turn) conductor. The secondary has many turns of small wire connected directly to a low-scale ammeter.

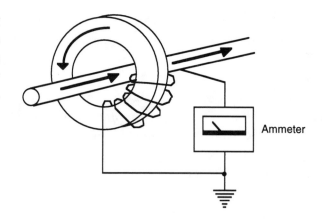

Ammeter

REMEMBER...

The **secondary** of a CT is rated at **5 amperes**.

K VA

VOLTAGE POTENTIAL TRANSFORMER

Usually circuits up to 600 volts can be measured directly with meters. However, higher voltages cause the meters to become very expensive in cost.

The potential transformer is used in metering of higher voltages. The primary winding is connected to the high-voltage and the secondary low-voltage winding usually wound for 120 volts.

The capacity of a potential transformer is relatively small as compared to a *power* transformer.

Potential transformers have ratings of 100 to 500 va.

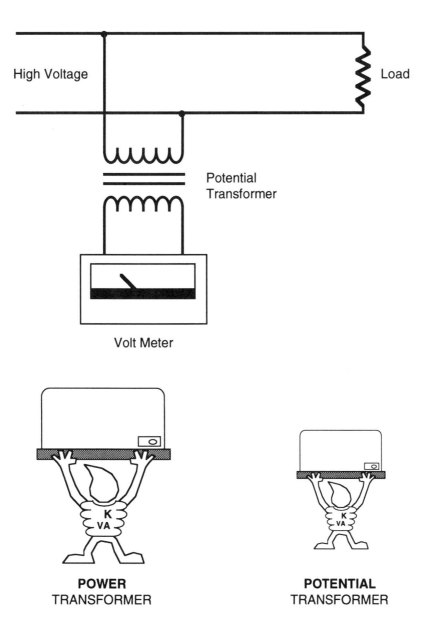

High Voltage

Load

Potential
Transformer

Volt Meter

POWER
TRANSFORMER

POTENTIAL
TRANSFORMER

WATTMETER

A wattmeter requires *both* voltage and current to measure power consumed. **W = E x I**

When measuring wattage in a high-voltage, high-current system, you would use both a voltage potential transformer and a current transformer.

Normally, two wattmeters are required to measure three-phase power, 3-wire system. But, a *balanced* three-phase load can be measured with *one* single-phase wattmeter.

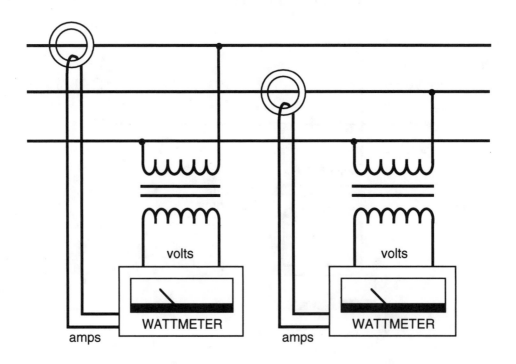

REMEMBER...

A **balanced** three-phase load can be measured with **one** single-phase wattmeter.

CONSTANT-CURRENT TRANSFORMER

Is used to supply a constant current for series street lighting either gaseous-discharge or incandescent lamps connected in series.

A moving coil *regulator* is usually used to feed a constant-current circuit.

The primary is normally 2400-10,000 volts, while the secondary is wound for the voltage required for the number of lights in series.

The constant-current value for the secondary in a series street-lighting system is normally 6.6 amps.

SMALL POWER TRANSFORMER

A constant-potential, self-air-cooled in smaller sizes 75, 150, 225 and 300 volt-amperes.

Used to step voltages down to 115v, 32v, 6v or so. Such as low voltage for landscape lighting.

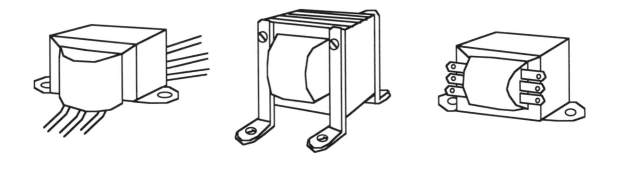

CONTROL and SIGNAL TRANSFORMER

Constant-potential, self-air-cooled used to step voltages down to supply control or signal circuits.

Used for the low voltage supply in motor control circuits. An even smaller capacity of this type transformer would be used to ring the door bell.

NEON-SIGN TRANSFORMER

Self-air-cooled mounted inside a sheet metal sign enclosure.

Primary voltage is normally 120 or 240 volt. The secondary varies from 2,000 to 15,000 volts. Current range 30 to 120 milliamperes to operate the gaseous-discharge neon lamps used in sign lighting.

POWER TRANSFORMER

The utility company uses the power transformer to raise generator voltage to a higher voltage for transmission to substations. This is a *step-up* transformer.

Voltages of 345,000 (345 kv) and higher are classified as *extra* high-voltage (EHV).

These higher voltages are used to transmit power over long distances.

The power transformer kva rating is determined by the power produced by the utility company generator. Usually the size limit due to physical size is 500 mkva (500 mega-volt-amps) 500,000,000 volt-amps.

SUBSTATION TRANSFORMER

Used by the utility company to *step-down* transmission voltages to intermediate voltage.

Example would be a transmission line voltage of 345,000v *stepped-down* to a lower voltage of 115,000v to 23,000v.

At a distribution substation the voltage is again *stepped-down* to 13,800v, 7200v or 6900v which is the primary voltage at the transformer on the pole or pad near your home.

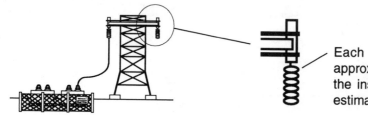

Each of these **insulators** can withstand approximately **10,000 volts**, and by counting the insulators in each string, a **reasonable** estimate of the applied voltage can be made.

DISTRIBUTION TRANSFORMER

This is the transformer on the pole or pad outside your home. This transformer is stepping-down the primary line voltage of say 7200 volts to 240 volts for the *loads to be served*.

A distribution transformer could be as large as 500 kva at 69,000 volts, but most distribution transformers are limited to 2,000 kva at 480/277 volt and 1000 kva 208/120 volt.

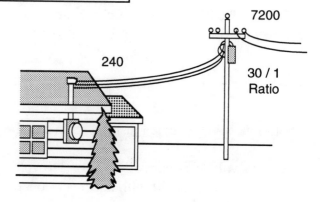

GLOSSARY of TERMS

AMPERE TURNS.

A unit of magnetomotive force which is obtained by multiplying the current in amperes by the number of turns in a winding.

ASKAREL and TRANSIL.

Special transformer oils that have a high insulating value. Oil is a good insulator.

AUTOTRANSFORMER.

A single winding transformer.

CORE.

The magnetic circuit upon which the windings are wound.

DIELECTRIC STRENGTH.

The maximum voltage that can be applied across the insulation safely.

EDDY CURRENT.

The flux changing in the iron circuit generates voltages in the laminated paths which set up a flow of currents at right angles to the flux path direction.

EFFICIENCY.

The ratio of output power to input power, expressed as a percentage.

EMF.

Electromotive force. Voltage.

HIGH-TENSION WINDING.

The winding which is rated for the higher voltage.

HYSTERESIS.

Hysteresis in an iron core means that the magnetic flux lags behind the magnetizing force that causes them.

INDUCED CURRENT.

A current that results in a closed conductor due to cutting the lines of magnetic force.

LAMINATIONS.

The iron core is constructed of thin sheets of soft iron which are insulated from each other. Laminations are used to reduce eddy-current loss.

LEAKAGE LINES.

The magnetic lines produced by the primary winding which do not pass through the secondary winding.

LENZ'S LAW.

It is necessary for 100,000,000 magnetic lines of force to be cut in one second to produce one volt. When a current is set up by an induced EMF due to the motion of a closed-circuit conductor, the direction of the current will be such that its magnetic field will oppose the motion.

MAGNETIC INDUCTION.

When lines of force are produced by an electric current.

MUTUAL INDUCTANCE.

Voltage applied to the primary winding induces a momentary voltage in the secondary winding. They are said to be coupled by mutual inductance. Mutual inductance exists between any two coupled coils.

NEUTRAL.

The neutral conductor carries the unbalanced current.

NO-LOAD CURRENT.

The current demand on the primary winding, when no current demand is made on the secondary winding.

PERMEABILITY.

The ease in which magnetic lines of force conduct with certain substances. Soft iron has good conductivity and does not have a high retentivity. Air is not a good conductor for magnetic lines, neither is copper.

POLARITY.

An electrical characteristic of EMF that determines the direction of current flow.

PRIMARY WINDING.

The winding which receives the energy, and it is not always the higher voltage winding.

RETENTIVITY.

The characteristic of retaining magnetic lines after the magnetizing cause has been removed. The harder the iron the greater the retentivity to the point of a so-called permanent magnet.

SATURATION POINT.

The relationship between flux density and the ampere turns for a magnetic material. When the saturation point is reached, it is no longer practical to increase the current or turns of a winding because the magnetic circuit is saturated.

SELF - INDUCTANCE.

Accounts for the production of a counter EMF in a winding that opposes the applied voltage.

SECONDARY TIES.

Is a circuit operating at 600 volts or less between phases which connects two power sources or power supply points.

SECONDARY WINDING.

The winding that delivers the power to the load.

TERTIARY WINDING.

A transformer with three windings rather than two.

TRANSFORMER RATIO.

The ratio between the voltage and the number of turns on the primary and secondary windings of a transformer is called the turns-ratio.

VOLTAGE TAPS.

The ratio adjuster provides a means of changing taps on the transformer to obtain a voltage ratio different from the standard ratio.

ABBREVATIONS

EHV = extra high-voltage
K   = kilo, one thousand (1000)
KV = kilo-volts
VA = volt-amperes
KVA = kilo-volt-amperes
MA = milliampere
MVA = mega-volt-amperes

SINGLE - PHASE

TRANSFORMER

CALCULATIONS

Transformers are divided into two groups, single-phase and three-phase.

In the single-phase transformer the fluxes are induced by the same voltage and are *in-phase* with each other.

In a three-phase transformer there are three flux paths, and each of the fluxes in these paths is displaced from the others by 120°. The three-phase transformer has three similar primary and secondary windings, one primary and one secondary winding for each phase. Therefore, the three-phase transformer is three single-phase transformers built on one magnetic circuit.

The term *in-phase* is the portion of a cycle or period through which the current or voltage has passed since going through zero value at the beginning of the cycle or period. Phase abbreviated = ø

This condition is called *true power* and is found in direct current (DC) circuits or in an AC circuit with only *resistance*.

When the circuit is in-phase: W = E x I   Watts (power) = volts times amps

Since many of the loads supplied by AC have induction, such as motors, transformers, etc. the current is *out-of-phase* with the voltage.

When out-of-phase this is called *apparent* power. Sometimes referred to as "floating power" because it does no work. Apparent power is called reactive power (volt-amperes).

Reactive volt-amperes (va) are sometimes referred to as reactive power, although they denote energy and not power. Reactive volt-amperes represent energy supplied to the circuit during part of a cycle and returned to the system during the time when the system inductance is discharging.

Power factor (PF) is the ratio between true power in *watts* and the apparent power in *volt-amps* (va).

Power Factor = watts/volt-amps   PF = W/VA

When *in-phase* the power factor is unity 1.0

DC or pure resistive AC circuits: W = E x I   WATTS = VOLTS x AMPS

Inductive AC circuit: VA = E x I   VOLT-AMPS = VOLTS x AMPS

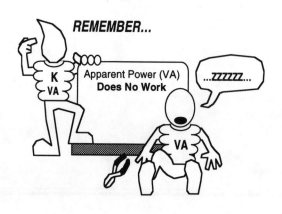

Transformers are sized in kva, smaller transformers in va.

$$kva = \frac{E \times I}{1000} \qquad va = E \times I$$

Example: What is the kva of a 1 ø transformer with a primary voltage of 240 and a current of 20 amps?

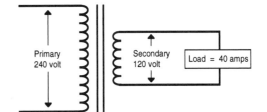

Primary 240 volt

Secondary 120 volt

Load = 40 amps

Solution: $kva = \frac{E \times I}{1000} \qquad \frac{240v \times 20a}{1000} = 4.8\ kva$

What is the va? $va = E \times I \qquad 240v \times 20a = 4800va$

Volt-amps in the secondary equal volt-amps in the primary at 100% efficiency.

Secondary voltage 120v x 40 amp load = 4800va same as primary

For exam calculations *assume* 100% efficiency unless otherwise specified in the calculation.

The primary winding is the *input* and the secondary winding is the *output*.

What is the turns-ratio of this transformer?

Turns Ratio = 240v/120v = 2   2/1 ratio step-down

Which has the larger size conductor the primary or the secondary?

The secondary would require the larger conductor for the 40 amperes of current. By the primary voltage being twice as high at 240v the current would be only half as much as the secondary at 1200 volts. Double the voltage, halve the current.

REMEMBER...

K VA

Volt-amps in the secondary **equal** volt-amps in the primary at 100% efficiency.

23

EFFICIENCY

With no moving parts transformers are *very efficient*. In general, the larger the transformer, the greater the efficiency. Some larger power transformer are 99% efficient.

Efficiency is always less than 100% due to the resistance (heat loss) in the windings.

The efficiency of a transformer is the ratio of the *output* to the *input*.

INPUT = PRIMARY WINDING   OUTPUT = SECONDARY WINDING

EFFICIENCY = OUTPUT/INPUT   INPUT = OUTPUT/EFF   OUTPUT = INPUT x EFF

When the efficiency goes down the input goes up for the output to remain the same.

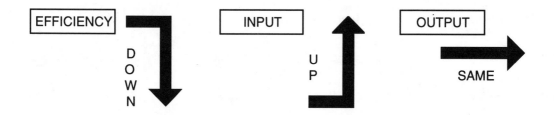

EFFICIENCY — DOWN   INPUT — UP   OUTPUT — SAME

This transformer has a secondary ouput of 4800 va. The primary draws 5000 va. What is the transformer efficiency?

Primary Input
5000va

Secondary
Output = 4800va

Solution: EFFICIENCY = $\dfrac{\text{OUTPUT}}{\text{INPUT}}$ = $\dfrac{4800 \text{ va}}{5000 \text{ va}}$ = .96 or 96%

The secondary output of this transformer is 12 amp at 96% efficiency. What is the primary current?

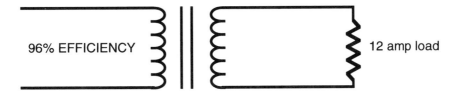

96% EFFICIENCY

12 amp load

Solution: INPUT = $\dfrac{\text{OUTPUT}}{\text{EFFICIENCY}}$ = $\dfrac{\text{12 amp}}{.96}$ = 12.5 amp INPUT

The input is 1500 va and the efficiency of the transformer is 96%. What is the output va?

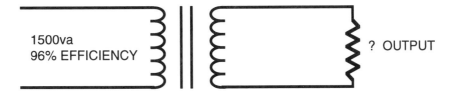

1500va
96% EFFICIENCY

? OUTPUT

Solution: OUTPUT = INPUT x EFFICIENCY = 1500va x .96 = 1440va OUTPUT

TRANSFORMER FORMULAS

E = volts   I = amps   P = primary   S = secondary

Ep = primary voltage
Es = secondary voltage
Ip = primary current
Is = secondary current

To find **primary voltage** when the current and secondary voltage are known:

$$Ep = \frac{Es \times Is}{Ip}$$

To find **primary current** when the secondary currents are known:

$$Ip = \frac{Es \times Is}{Ep}$$

To find **secondary voltage** when the current and primary voltage are known:

$$Es = \frac{Ep \times Ip}{Is}$$

To find **secondary current** when the voltages and primary current are known:

$$Is = \frac{Ep \times Ip}{Es}$$

$$Ep = \frac{Es \times Is}{Ip}$$

What is the primary voltage of this transformer?

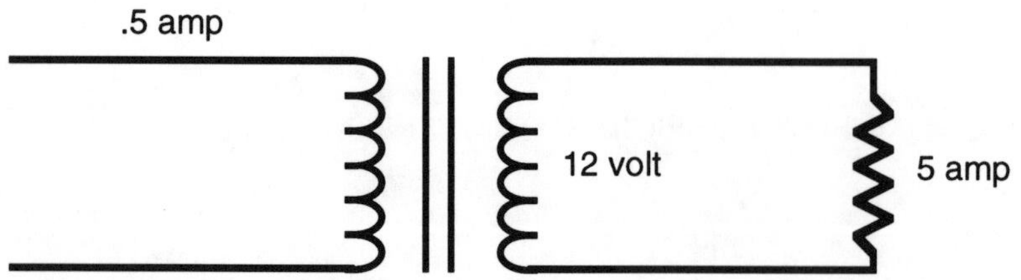

.5 amp

12 volt   5 amp

Solution: Primary voltage   = $\dfrac{12v \times 5\ amp}{.5\ amp}$ =   120 primary voltage

---

$Ip = \dfrac{Es \ x \ Is}{Ep}$

What is the primary current of the following transformer?

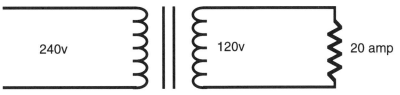

Solution: Primary current   = $\dfrac{120v \ x \ 20 \ amp}{240v}$ =   10 amp primary current

$Es = \dfrac{Ep \ x \ Ip}{Is}$

What is the secondary voltage of the following transformer?

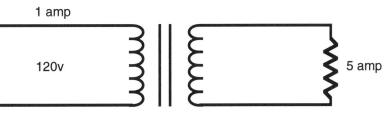

Solution: Secondary voltage   = $\dfrac{120v \ x \ 1 \ amp}{5 \ amp}$ =   24 secondary voltage

$Is = \dfrac{Ep \ x \ Ip}{Es}$

What is the secondary current of the following transformer?

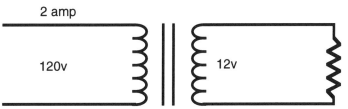

Solution: Secondary current   = $\dfrac{120v \ x \ 2 \ amp}{12v}$ =   20 amp secondary current

POWER FACTOR

Power factor (PF) is the ratio between *true* power in watts (w) and *apparent* power in volt-amps (va).

$$PF = w/va$$

When the circuit is "in-phase" we have true power (watts). Since most of the loads supplied by AC have induction, such as motors, transformers, etc., the current is out-of-phase with the voltage. The difference between "in-phase" (watts) and "out-of-phase" (va) is called the power factor (PF).

Unity ("in-phase") = 1.0   Power factor ("out-of-phase") = **less** than 1.0

True power (watts) is the power that is actually consumed (used up). But we *size* electrical equipment and conductors by the apparent power (va).

Your electric bill is calculated by kwh (kilo-**watt**-hour). The transformer is sized by kva (kilo-**volt-amperes**).

Power factor measures the angular relationship between voltage and current. Power factor causes a greater current to occur on the input.

Example: A 200 amp load at 240v 1ø = 48000va or 48 kva. If the power factor was perfect (unity 1.0), this would be equivalent to 48 kw. At 7¢ per kilo-watt-hour x 48 kw, the power company would receive $3.36 per hour for the power. But with a power factor **less** than 100% ("out-of-phase") say 75% power factor. You still have 48 kva but only 36 kw (48 x 75%), and at 7¢ per kilo-watt-hour, the power company now receives $2.52 per hour (36 x .07) for the power. Since the kva determines the wire size, transformer and generator size, the power company must furnish wires just as large for $3.36 per hour as for when they are paid only $2.52 per hour, and they still tie up just as much generator and transformer capacity.

This is why power companies add a penalty to the monthly bill for industrial users with low power factors. The power factor can be corrected by means of capacitors or synchronous motors.

The advantages of unity power factor (1.0):

• The power loss (I^2R) in the line is reduced, efficiency of transmission is increased
• Less voltage drop
• Higher efficiency and operating performance of AC generators and transformers supply the circuits

$\text{Cos } \theta$ = Power factor (cosine of theta) $PF = \text{Cos } \theta = R/Z$

$$PF = w/va \qquad W = va \times PF \qquad I = \frac{w}{E \times PF} \qquad E = \frac{w}{I \times PF} \qquad VA = W/PF$$

SINGLE - PHASE

LOAD BALANCING

The Edison 3-wire system has two line wires and a *neutral* conductor.

With the 3-wire 230/115 volt system there are two voltage levels. 230 volt single-phase loads are connected line to line. 115 volt loads are connected line to neutral.

An example of this system is the service to your residence. The transformer on the pole or pad serving your residence may have a primary voltage of 6900v and a secondary voltage to your house of 230 volts. A ratio of 30 to 1.

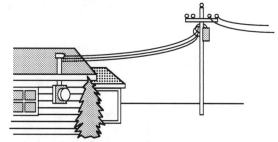

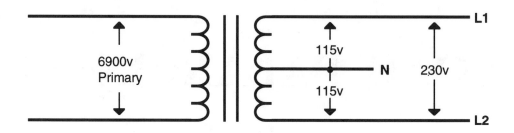

The *center* of the secondary winding is tapped with the neutral conductor and grounded.

The neutral conductor will carry the *unbalanced* current from L1 and L2.

When the designer has calculated the size of the transformer and conductors it is the responsibility of the electrician connecting the loads in the panel to maintain this *balancing* as designed.

Any additions or changes should be calculated and balanced to prevent overloading of the transformer and the conductors.

In a 2-wire, 120v circuit, the grounded conductor (identified white in color) is *not* a neutral. In a 2-wire circuit the grounded (white) conductor carries the same amount of current as the ungrounded (hot) conductor carries.

REMEMBER...

A **neutral** conductor carries the **unbalanced** current; you must have a **3-wire** circuit to have a **neutral** conductor.

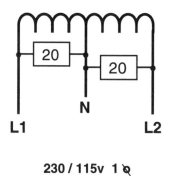

230 / 115v 1 ∅

An example of a balanced circuit, the two 20 amp, 115 volt loads are connected *balanced*.

Since the *neutral* carries only the *unbalanced current*, with both loads on it will carry *zero current* in this example of a balanced circuit.

The neutral would be required to carry the *maximum unbalanced current.*

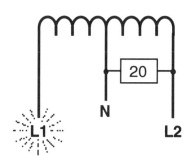

If the load on L1 was shut off, the neutral conductor would carry the 20 amp load connected between L2 and the neutral, so in this circuit the *maximum* unbalance is 20 amperes.

Next is an example of the electrician connecting the two 20 amp, 115 volt loads incorrectly.

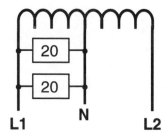

With both of the 20 amp loads connected from L1 to neutral unbalanced, with both loads turned on, the neutral conductor would have to carry 40 amps maximum instead of the 20 amp maximum in a balanced circuit. Twice the current, the neutral conductor would have to be sized larger to carry this extra current.

NOW YOU CAN
SEE THE
IMPORTANCE
of
PROPER LOAD
BALANCING

An example of two unequal loads, one 10 amp 115v, and one 20 amp 115v:

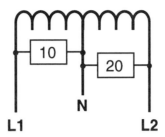

With both loads on, the neutral conductor would carry the unbalanced current of 10 amperes (20 amp - 10 amp = 10 amp). The maximum current the neutral conductor would carry is 20 amperes when L1 to neutral load of 10 amps is shut off, then the neutral would carry the L2 to neutral load of 20 amps.

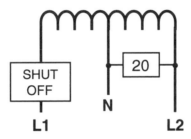

With the L1 load shut off, the neutral conductor would carry 20 amps.

Balance the following loads as closely as possible. 3-wire, 1 ø, 230/115v.

1 - 5000 watt 1 ø, 230v load
2 - 3000 watt 1 ø, 230v loads
6 - 400 watt 1 ø, 115v loads

1 ø 230 / 115v

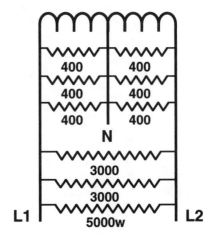

The 5000 watt load is 230v so it would connect between L1 and L2. The two 3000 watt loads are also 230v so they connect between L1 and L2.

The six 400 watt loads are 115v so they connect from line to neutral and are balanced as closely as possible. In this case we can obtain a balance by connecting three 400 watt loads on line one to neutral and the other three 400 watt loads line two to neutral.

If the loads are not 115 volt they do not connect to the neutral. Only the 115 volt loads connect to the neutral.

Calculate the current in each of the conductors L1, L2, and the neutral with all loads on:

CURRENT IN LINE ONE:

Line one would carry half of the 5000 watt load   = 2500 watt
Line one would carry half of the two 3000 watt loads   = 3000 watt
Line one would carry three 400 watt loads   = 1200 watt
                                                                                          6700 watt

$\dfrac{6700 \text{ watt}}{115 \text{ volt}}$ = 58.26 amperes current in line one

CURRENT IN LINE TWO:

Line two would carry half of the 5000 watt load	= 2500 watt
Line two would carry half of the two 3000 watt loads	= 3000 watt
Line two would carry three 400 watt loads	= 1200 watt
	6700 watt

$$\frac{6700 \text{ watt}}{115 \text{ volt}} = 58.26 \text{ amperes current in line two}$$

The current in the neutral conductor *with all loads on would be zero*.

The *maximum current* the neutral conductor would carry in this circuit connected in a balance would be when all the loads on either line one or line two were shut off.

Example: With line one loads shut off the neutral would carry the loads connected line two to neutral.

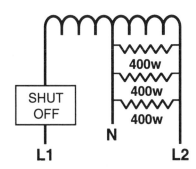

$$\frac{1200\text{w}}{115\text{v}} = 10.43 \text{ AMPERES } \textit{MAXIMUM} \text{ NEUTRAL CURRENT!}$$

The electrician can perfectly balance these loads as they are connected in the panel. But, it's almost impossible to maintain a perfect balance as all loads will not always all be on, nor will all the loads always be off.

So as the 400 watt, 115 volt loads are switched on and off the current flowing in the neutral conductor varies from zero current with all loads on to a maximum current of 10.43 amps for three loads on one line at the same time with all loads off on the other line.

Total power required by all the loads in EXAMPLE ONE would be:

1 - 5000 watt = 5000 watt
2 - 3000 watt = 6000 watt
6 - 400 watt = 2400 watt
                13,400 watt total power

$$\text{TOTAL CURRENT} = \frac{\text{TOTAL WATTAGE}}{\text{TOTAL VOLTAGE}} = \frac{13,400 \text{ watt}}{230 \text{ volt}} = 58.26 \text{ amperes current}$$

At 100% efficiency and PF 1.0 the primary would also be 13,400.

If the primary voltage is 6900 volts the primary current would be:

$$\frac{13,400}{6900} = 1.94 \text{ amperes primary current}$$

6900/230 = 30/1 ratio

Primary current = 1.94 amperes x 30 ratio = 58.2 amperes secondary current.

Now you can see why the primary conductors are smaller in size than the secondary conductors.

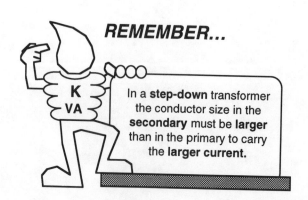

REMEMBER...

In a **step-down** transformer the conductor size in the **secondary** must be **larger** than in the primary to carry the **larger current.**

An unbalanced calculation:

1 - 6000 watt 1 ø 230 volt load
1 - 4000 watt 1 ø 230 volt load
5 - 1500 watt 1 ø 115 volt loads

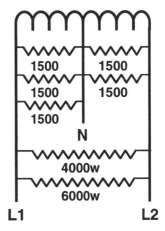

The 6000 watt load is 230 volt, it would connect between line one and line two. The 4000 watt load is also 230 volt so it would also connect between line one and line two.

The five 1500 watt loads are 115 volt, they would connect line to neutral. They are balanced as closely as possible. In this case connect three 1500 watt loads from line one to neutral and two 1500 watt loads from line two to neutral.

Calculate the current in each of the conductors L1, L2, and the neutral with all loads on:

CURRENT IN LINE ONE:

Line one would carry half of the 6000 watt load   = 3000 watt
Line one would carry half of the 4000 watt load   = 2000 watt
Line one would carry three 1500 watt loads   = 4500 watt
                                  9500 watt

$$\frac{9500 \text{ watt}}{115 \text{ volt}} = 82.6 \text{ amperes current in line one}$$

CURRENT IN LINE TWO:

Line two would carry half of the 6000 watt load   = 3000 watt
Line two would carry half of the 4000 watt load   = 2000 watt
Line two would carry two 1500 watt loads   = 3000 watt
                                                     8000 watt

$\dfrac{8000 \text{ watt}}{115 \text{ volt}}$ = 69.56 amperes current in line two

Current flowing in the neutral with all the loads on would be 13.04 amperes, the unbalanced current. 1500 watt the unbalanced divided by 115 volt = 13.04 amperes.

The maximum current the neutral conductor would carry connected as shown, would be when the line two loads are both off, the three loads line one to neutral would be the maximum unbalance 4500w/115v = 39.13 amperes maximum current.

Total power required by all the loads in EXAMPLE 2 would be:

1 - 6000 watt = 6000 watt
1 - 4000 watt = 4000 watt
5 - 1500 watt = 7500 watt
                      17,500 watt total power

$$\text{TOTAL CURRENT} = \frac{\text{TOTAL WATTAGE}}{\text{TOTAL VOLTAGE}} = \frac{17,500 \text{ watt}}{230 \text{ volt}} = 76.08 \text{ amperes current}$$

LOAD BALANCING EXAMPLE 3

Determining the neutral current in an unbalanced single-phase 3-wire, 230/115 volt system.

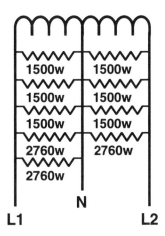

L1	L2
1500 watt	1500 watt
1500 watt	1500 watt
1500 watt	1500 watt
2760 watt	2760 watt
2760 watt	------------
10,020 watt	**7,260 watt**

Total power would be 10,020 watt plus 7,260 watt = 17,280 watt

$$\frac{17,280 \text{ watt}}{230 \text{ volt}} = 75.13 \text{ amperes current}$$

By assuming the loads to balanced and dividing the total power by the line voltage the total current equals 75.13 amperes. But, for more exact results for an *unbalanced* system, calculate the neutral load by the largest line to neutral total 10,020 watts divided by the line to neutral voltage of 115, equals 87.13 amperes or 12 amperes higher.

So the maximum neutral current as connected would be if the line two loads were off. The line one to neutral current would be the maximum 87.13 amperes.

CALCULATING THE NEUTRAL LOAD

When a 3-wire 230/115 volt circuit serves a total load that is balanced from each hot leg to neutral - that is, half the total load is connected from one hot leg to neutral and the other half of the total load from the other hot leg to neutral - the condition of maximum unbalance occurs when all the load fed by one hot leg is operating and all the load fed by the other hot leg is off. Under that condition, the neutral current and hot leg current are equal to half the total watts divided by 115 volts (half the voltage between hot legs).

But that current is exactly the same as the current that results from dividing the total load (connected hot leg to hot leg) by 230 volt (which is twice the voltage from hot leg to neutral) because of this relationship, it is easy to determine the neutral current by simply calculating hot-leg load (the total load from hot leg to hot leg) and dividing by 230 volt.

COMPUTATION ACCURACY

When 115 volt loads are exactly balanced on a 230/115 volt circuit, it is correct to add the load in watts and then divide by 230 volts to determine the line and neutral load in amperes. The neutral load represents the maximum unbalanced load at 115 volts. For balanced 115 volt loads, the neutral is one-half the total load; therefore, dividing the total load in watts by 230 volt yields the correct value for the neutral in amperes.

If a 115 volt load is *not balanced* by an identical load, the neutral load should *not* be determined by dividing by the 230 volt circuit voltage if an *exact* result is desired.

Standard practice allows the neutral load in amperes to be determined by summing the loads that require a neutral and dividing the total watts by 230 volts. It should be remembered that this method is slightly *inaccurate* when *unbalanced* loads are present on 3-wire circuits.

EXAMPLE 3 on page 39 shows the **exact** calculation at 87.13 amperes.

The following example shows two 10 amp, 115 volt loads connected in a balance between L1 and L2.

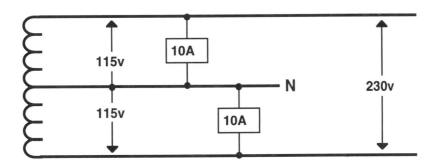

With both loads on and balanced, the circuit is actually 230 volts with the two 10 amp loads in series.

Since the loads have equal resistance the voltage of 230 would divide, 115 volts to each load. This is a balanced circuit.

The neutral conductor carries zero current in a balanced circuit.

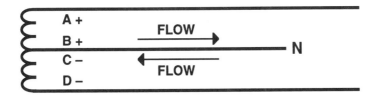

Each of the conductors B and C carries 10 amps, but note the direction of the arrows. The flow of current in conductor B at any *instant* is in an *opposite direction* of conductor C. So at any given instant, the single neutral conductor is said to be carrying 10 amps in one direction and 10 amps in the opposite direction. The two cancel each other and the flow of current in the neutral conductor is *zero*.

SINGLE-PHASE

EXAMS

SINGLE-PHASE EXAM 1

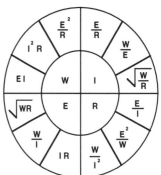

A single-phase transformer has a primary voltage of 120, a secondary voltage of 24. The load is 240va.

1. What is the turns ratio of this transformer?

(a) 2/1   (b) 1/2   (c) 5/1   (d) 1/5

2. What is the secondary current?

(a) 10 amp   (b) 5 amp   (c) 2 amp   (d) 15 amp

3. What is the primary current?

(a) 10 amp   (b) 5 amp   (c) 2 amp   (d) 15 amp

4. What is the va input?

(a) 2 amp   (b) 10 amp   (c) 120 va   (d) 240 va

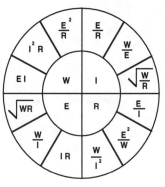

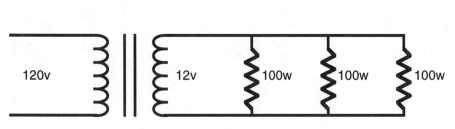

1. A pool has three 100 watt, 12 volt lights, the load is ____ watts.

(a) 300   (b) 200   (c) 100   (d) 50

2. The current in the primary is ____ amperes.

(a) 12/120   (b) 25   (c) 2.5   (d) .25

3. The turns ratio of the transformer is ____.

(a) 12/120   (b) 2/1   (c) 10/1   (d) 1/10

4. The current in the secondary is ____ amperes.

(a) 12/300   (b) 25   (c) 2.5   (d) .25

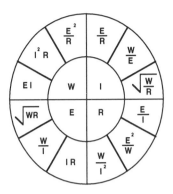

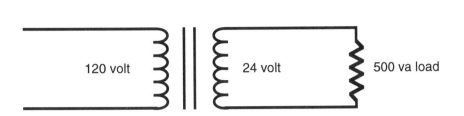

1. What is the turns ratio?

(a) 2/1   (b) 4/1   (c) 5/1   (d) 10/1

2. What is the secondary current?

(a) 208.3 amp   (b) 50 amp   (c) 20.83 amp   (d) 28 amp

3. What is the primary current?

(a) 50 amp   (b) 4.16 amp   (c) 42 amp   (d) 20.83 amp

4. What is the input va?

(a) 500 va   (b) 250 va   (c) 100 va   (d) 750 va

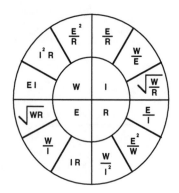

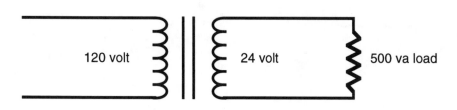

TRANSFORMER IS 96% EFFICIENT

1. What is the secondary current?

(a) 21.7 amp   (b) 20.8 amp   (c) 50 amp   (d) none of these

2. What is the input volt-amps?

(a) 500 va   (b) 521 va   (c) 750 va   (d) 485 va

3. What is the primary current?

(a) 20.8 amp   (b) 4.17 amp   (c) 4.34 amp   (d) 50 amp

4. What is the primary kva?

(a) 500   (b) 20.8   (c) 521   (d) .521

SINGLE-PHASE POWER FACTOR EXAM 5

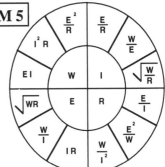

1. What is the power factor of a 6500 watt load connected to a 240 volt single-phase source that draws 33 amps?

(a) 1.21%   (b) 82%   (c) 68%   (d) 1.04%

2. At what power factor is a single-phase system operating at when the wattmeter reads 2000 watts, the voltmeter reads 230 volts, and the ammeter reads 10.1 amps?

(a) 86%   (b) 1.16%   (c) 72%   (d) 89%

3. How much current does a 15 kw load take from a 240 volt single-phase system at 84% power factor?

(a) .013 amp   (b) 52.5 amp   (c) 74.4 amp   (d) 42.8 amp

4. What is the true power of a 10,000 va load, power factor is 85%, single-phase 230 volt?

(a) 11,764 watt   (b) 8500 watt   (c) 1955 watt   (d) 2500 watt

5. What is the power factor of a single-phase load, input va = 25,000, output watts = 21,500?

(a) 1.16%   (b) 86%   (c) 94.5%   (d) 79%

6. What voltage will be required to supply a 5000 watt load with 25 amperes at 85% power factor?

(a) 208v   (b) 115v   (c) 220v   (d) 235v

7. What is the efficiency of a single-phase transformer that has an input of 92 amps on 240 volts with with an output of 20 kw?

(a) 98%   (b) .905   (c) 62%   (d) 94%

8. What is the current to a 15 kw rated load with a power factor of 86%, 240v single-phase?

(a) 62.5 amp   (b) 72.67 amp   (c) 74.9 amp   (d) 86 amp

SINGLE-PHASE EXAM 6

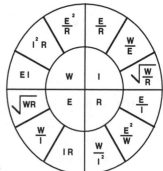

1. A transformer with a primary of 120 volt and a secondary of 15 volt, what is the turns ratio of this transformer?

(a) 2/1   (b) 4/1   (c) 6/1   (d) 8/1

2. If the load on the 24 volt secondary of a transformer is 100 watts, the secondary current is _____ amperes.

(a) 41.6   (b) 4.16   (c) .416   (d) .83

3. A transformer has an output load of 3550 va at 97% efficiency, what is the input va?

(a) 3660 va   (b) 3444 va   (c) 3608 va   (d) 3395 va

4. A single-phase isolation transformer has a load of 240va, the transformer is 120/12 volt. What is the primary current when the load is on?

(a) 2 amp   (b) 20 amp   (c) 2.4 amp   (d) 24 amp

5. A voltage transformer has a turns ratio of 5/1. The wire on the _____ is larger.

(a) primary   (b) secondary   (c) tertiary   (d) windings are the same

6. A single-phase 240/120v secondary has five 120v 1200 watt loads balanced as closely as possible, with all loads on what is the current flowing in the neutral?

(a) 10 amp   (b) 30 amp   (c) 5 amp   (d) 20 amp

7. What is the power factor of a 7500 watt load connected to a 230 volt single-phase source that draws 38 amperes?

(a) 86%   (b) 1.16%   (c) 92%   (d) 79%

8. A transformer with a 120 volt primary, 12 volt secondary, load of five 60 watt bulbs, what is the primary current?

(a) 25 amp   (b) 2.5 amp   (c) 5 amp   (d) 0.5 amp

48

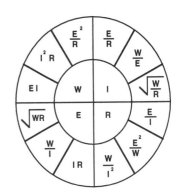

1. A transformer with a primary voltage of 480, secondary voltage 240/120. What is the turns ratio of this transformer?

(a) 4/1   (b) 1/4   (c) 2/1   (d) 1/2

2. How much current does a 12 kw load take from a 230 volt single-phase system at 88% power factor?

(a) 52.17 amp   (b) 45.9 amp   (c) 59.28 amp   (d) 65 amp

3. Six 1500 watt 115 volt loads are balanced on the secondary of a 230/115v single-phase transformer. With all loads on, what is the current flowing in the neutral?

(a) 39.1 amp   (b) 78.2 amp   (c) 19.56 amp   (d) 0

4. A transformer has a secondary load of 500va, secondary voltage is 48, primary voltage is 240, transformer efficiency is 96%. What is the primary current?

(a) 2.08 amp   (b) 21.7 amp   (c) 2.17 amp   (d) 10.85 amp

5. The primary of a transformer draws 5 amperes at 7200 volts. A reading at the secondary shows 140 amperes at 240 volts. What is the efficiency of this transformer?

(a) 1.07%   (b) 93%   (c) 88%   (d) 98%

6. A single-phase transformer with a secondary voltage of 120 has a load of 30 amp. The transformer is 98% efficient. What is the primary kva at 240v?

(a) 3673   (b) 3600   (c) 3.67   (d) 4800

7. A singe-phase transformer with a 240/120v secondary has the following loads:

2 - 2500 watt 1 ø 240v
1 - 1200 watt 1 ø 120v
1 - 3000 watt 1 ø 240v
4 - 1500 watt 1 ø 120v

With the loads balanced as closely as possible, what is the maximum neutral current?

(a) 35 amp   (b) 25 amp   (c) 57.9 amp   (d) 66.08 amp

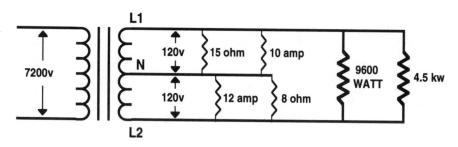

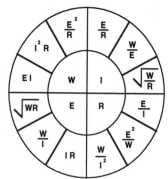

1. What is the turns ratio of this transformer?

(a) 60/1 (b) 4/1 (c) 30/1 (d) 1/2

2. With all loads on, what is the current flowing in L1?

(a) 75.76 amp (b) 162.5 amp (c) 76.75 amp (d) 150 amp

3. With all loads on, what is the current flowing in L2?

(a) 85.75 amp (b) 87.55 amp (c) 162.5 amp (d) 150 amp

4. With all loads on, what is the current flowing in the neutral?

(a) 162.5 amp (b) 85.75 amp (c) 76.75 amp (d) 9 amp

5. What would be the maximum current the neutral conductor would carry?

(a) 9 amp (b) 162.5 amp (c) 85.75 amp (d) 27 amp

6. What is the kva load on this transformer with all loads on?

(a) 19500 (b) 10290 (c) 9210 (d) 19.5

7. What is the current flowing in the secondary with all loads on?

(a) 85.75 amp (b) 76.75 amp (c) 162.5 amp (d) 81.25 amp

8. What is the primary current with all loads on?

(a) 2.71 amp (b) 81.25 amp (c) 162.5 amp (d) 27 amp

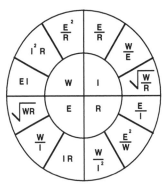

1. If a circuit has a load of 14 amps at 220 volt, what is the va if the power factor is .95?

(a) 3080   (b) 2926   (c) 3630   (d) 3242

2. A circuit has a voltage of 220 AC and a load of 14 amp. A wattmeter indicates 2900 watt load. What is the power factor of this circuit?

(a) .94   (b) 1.94   (c) 13.1   (d) 45571

3. The kva input is most nearly _____ if the load of the transformer is 500 watt.

(a) 1500   (b) 500   (c) .5   (d) 50

4. A single-phase transformer has a primary of 480v and a secondary of 240/120v, with a neutral center tapped in the secondary. If the primary is rated 45 kva the load that can be attached to the secondary would be a maximum of _____, assuming 100% efficiency.

(a) 22.5 kva   (b) 45 kva   (c) 22 1/2 watts   (d) 45 watts

5. Which of the following is true?

(a) a single-phase 3-wire system provides three 120v conductors to ground
(b) current is high with high-voltage and lower with low-voltage systems
(c) autotransformers have a single-winding that is common to both the primary & secondary
(d) when the primary winding has less turns than the secondary the input voltage would be higher

6. What is the amperage for a 50 kva single-phase load with a primary voltage of 480?

(a) 52 amp   (b) 104 amp   (c) 208 amp   (d) 240 amp

7. A single-phase transformer 480v primary, 240/120v secondary with a center tap neutral. The transformer is rated 48 kva, a 240v load is 24 kva, a balanced 120v load would need a neutral of _____ amps.

(a) 50   (b) 100   (c) 150   (d) 200

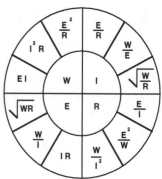

1. 480/240/120v single-phase center tap transformer, what is the turns ratio?

(a) 4 to 1   (b) 1 to 4   (c) 2 to 1   (d) 1 to 2

2. The primary of a transformer draws 6 amps at 7200 volts, the secondary shows a reading of 165 amps at 240 volts. What is the efficiency?

(a) 21.6%   (b) 30%   (c) 72%   (d) 92%

3. If three 1900 watt lighting circuits and two 2 hp single-phase pumps are balanced as closely as possible on a 3-wire, 240/120v system, and all loads are on, the neutral will be carrying approximately ____ amps if one motor is on each hot wire.

(a) 0   (b) 16   (c) 40   (d) 50

4. Which of the following is false?

(a) secondary windings with less turns than the primary step voltages down
(b) the secondary winding is the output winding
(c) with unbalanced current on the neutral, there is zero current flow on the grounded conductor
(d) kva is derived by adding all the va loads together and dividing by 1000

5. A single-phase 240/120v transformer has a secondary load of 24 amperes. The transformer is 98% efficient. What is the primary kva at 240 volts?

(a) 5760   (b) 2938.7   (c) 294   (d) 2.94

6. The maximum current on a neutral with either L1 or L2 on is ____ amps if the load on L1 is 9000 watts and the load on L2 is 8500 watts. The voltage is 240/120.

(a) 40   (b) 120   (c) 27   (d) 75

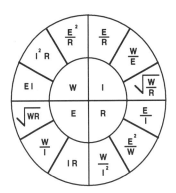

1. What is the power factor of a 600 watt load connected to a 120 volt single-phase source that draws 6 amps?

(a) 1.96   (b) .83   (c) 29.98   (d) .43

2. Which of the following is true?

(a) secondary windings with more turns than the primary winding step voltages down
(b) the current for the secondary is determined by dividing the kva by the primary voltage
(c) the core of transformers is magnetized by the current flow through the primary winding
(d) the input va will be lower than the output va if the efficiency is less than 100%

3. Which of the following is false?

(a) connecting transformer windings in series will give higher voltages
(b) in a three-wire circuit the neutral carries the balanced current
(c) if the primary winding has less turns than the secondary, the secondary output will be higher
(d) the input va will be higher than the output va if the efficiency is less than 100%

4. What is the applied voltage on this winding shown below?

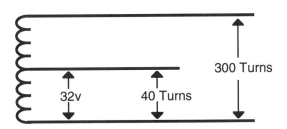

300 Turns

32v   40 Turns

(a) 24v   (b) 40v   (c) 120v   (d) 240v

5. What is the current to a 25 kw rated load with a power factor of 82%, 240v single-phase?

(a) 127 amp   (b) 1.27 amp   (c) 85.4 amp   (d) 12.7 amp

6. A transformer with a 120v primary, 12v secondary, used for landscape lighting has a load of ten 12 watt lights, what is the primary current?

(a) 10 amp   (b) 5 amp   (c) .1 amp   (d) 1 amp

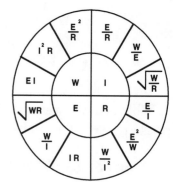

1. A single-phase transformer has an input of 5263va, the efficiency is 95%, the output would be approximately _____ watts.

(a) 5540 (b) 5000 (c) 6000 (d) 7500

2. The following loads are connected to a 240/120v single-phase secondary and balanced as well as possible:

3 - 1900 watt lighting loads 120v
2 - 2 hp single-phase motors 120v

With all loads on, what is the current in the neutral conductor?

(a) 15.8 amp (b) 32.1 amp (c) .5 amp (d) 31.6 amp

3. The following loads are connected to a single-phase transformer 240/120v secondary with a center-tap neutral:

1 - 5000w water heater 240v
4 - 1200w lighting loads 120v
1 - 6000w space heat 240v
5 - 1200w appliance loads 120v

With all loads turned on and balanced as closely as possible, what is the current flowing in the neutral?

(a) 10 amp (b) 181.6 amp (c) 90.8 amp (d) 100 amp

4. What is the amperage for a 25 kva single-phase load with a secondary voltage of 240?

(a) 52 amp (b) 104 amp (c) 208 amp (d) 240 amp

5. What voltage will be required to supply a 10,000 watt load with 50 amperes at 83.5% power factor?

(a) 115v (b) 120v (c) 230v (d) 240v

6. If the load on the secondary of a 120/24v transformer is 150 watt, the primary current would be _____ amperes.

(a) 6.25 (b) 12.5 (c) 1.25 (d) 31.25

THREE - PHASE

TRANSFORMER

CALCULATIONS

For power applications single-phase systems are inefficient and unsatisfactory.

Comparing a single-phase system to a three-phase system would be like comparing a single-cylinder gasoline engine to a three-cylinder engine.

By the use of three conductors a three-phase system can provide 173% more power than the two conductors of a single-phase system.

A three-phase system has three sources of power with a time interval between each source. The windings of the generator are located 120° apart. As the windings rotate through the generator magnetic field the induced voltages will be 120° apart, one-third of a complete 360° cycle.

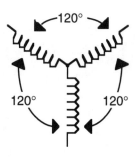

A three-phase system is three single-phase voltages connected together.

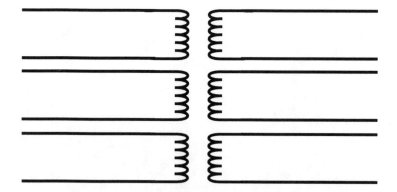

But rather than having six conductors they can be connected in a *delta* or *wye* configuration and reduce the six line wires to three.

The delta connection is shown as the triangle $\triangle$.

The delta has the three windings connected in a closed circuit. The ends of the windings connect together in the proper polarity.

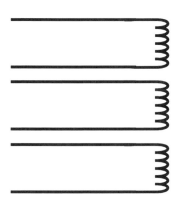

DELTA

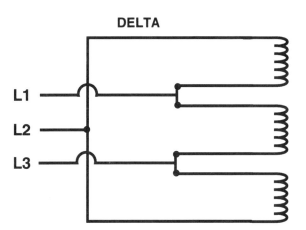

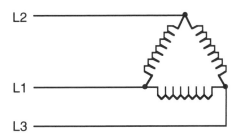

The wye, or sometimes called the star is shown as Y.

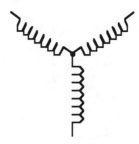

The wye is the most commonly used way of connecting the three single-phase windings, all three windings connect together at one point.

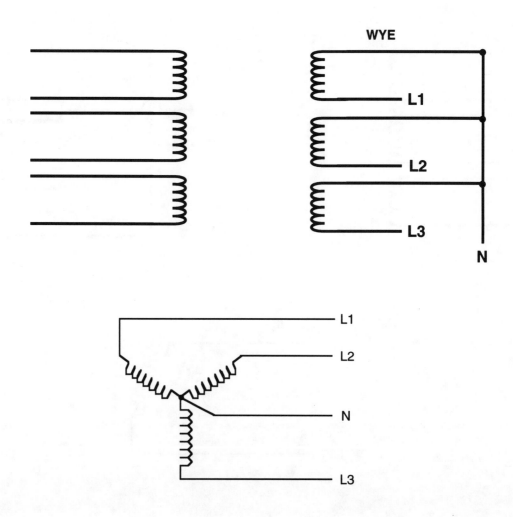

The four possible combinations of transformer windings are:

- wye to wye connected
- wye to delta connected
- delta to delta connected
- delta to wye connected

The **wye to wye** has its advantages and its disadvantages, but the disadvantages predominate and have limited its use. The disadvantages include disturbances from harmonic voltages and currents, also loads that are unbalanced cannot be carried on the secondary unless a primary neutral is provided. Wye to wye connections are not often used in interior electrical wiring. The wye to wye is found most often on high-voltage transmission.

Wye to delta is used to step down high transmission line voltages.

Delta to delta is mainly used to supply industrial loads by stepping down 2400 volt three-phase to a 240 volt three-phase, 3-wire service.

Delta to wye is the most popular connection for distribution service where a four-wire secondary distribution circuit is desired.

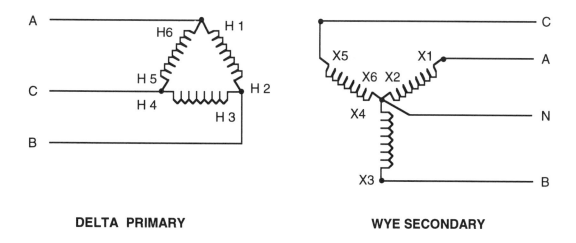

DELTA PRIMARY   **WYE SECONDARY**

REMEMBER...

The most common connection: **Delta-Wye**

59

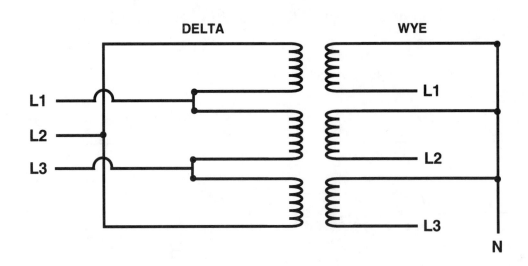

DELTA   WYE

L1

L2

L3

L1

L2

L3

N

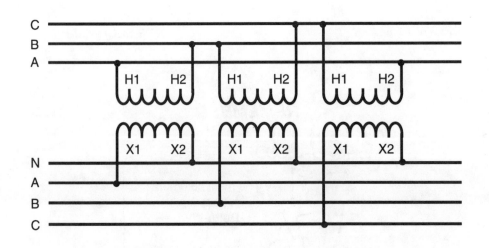

Delta-wye 480/208/120 volt three-phase system. By connecting the three single-phase transformers in this configuration let's check the voltage and current relationship that is now established.

Wye secondary 208/120 volt three-phase, the three phases are marked A, B, C:

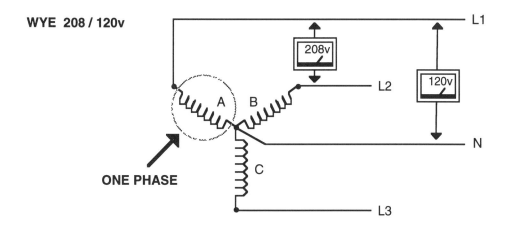

To check the *phase* voltage you would place the voltmeter from L1 to neutral, this would read the the voltage across *one phase* (or winding), the voltage would read 120. Thus the phase voltage is 120.

To check the *line* voltage you would place the voltmeter from L1 to L2, now you are reading the voltage across *two phases* (A and B), the voltmeter would read 208 volts. The line voltage is 208.

Phase voltage times 1.732 equals line voltage in a wye (120v x 1.732 = 208v).

Line voltage divided by 1.732 equals phase voltage (208v/1.732 = 120v).

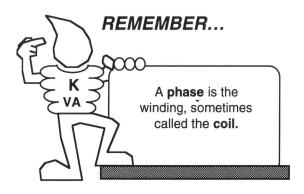

REMEMBER...

A **phase** is the winding, sometimes called the **coil.**

Reading the line voltage from L1 to L2 you are actually reading across *two* phases. The voltage of one phase is 120, but the voltage of the two phases is **not** 240. The voltages of the two phases are 120° apart, so the line voltage equals the phase voltage times 1.732. (120v x 1.732 = 208v)

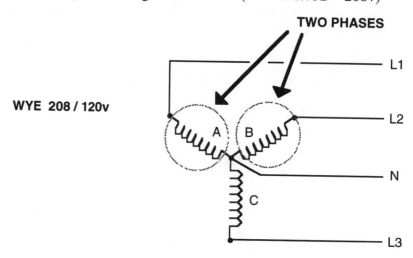

WYE 208 / 120v

Let's check the current relationship in this wye secondary. If the phase current is 10 amps, the line current is also 10 amps. The current is the *same* in a wye because the phase is connected in series with the line.

Phase amps equal line amps in a wye.

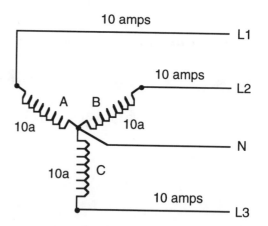

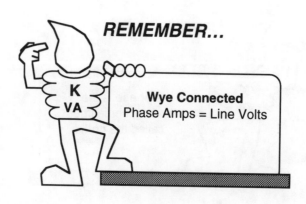

REMEMBER...

Wye Connected
Phase Amps = Line Volts

WYE CONNECTED

PHASE AMPS = LINE AMPS

PHASE VOLTS x 1.732 = LINE VOLTS

——— OR ———

$$\frac{\text{LINE VOLTS}}{1.732} = \text{PHASE VOLTS}$$

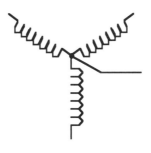

Delta primary 480 volt. Let's check the delta voltage and current relationship.

DELTA 480v PRIMARY

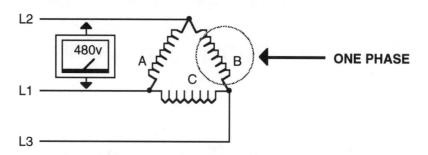

To check the *phase* voltage you would place the voltmeter across one phase which would be L1 to L2, the phase voltage is 480 volts. To check the *line* voltage you would *also* place the voltmeter across L1 to L2, the line voltage is also 480.

Phase volts equal line volts in a delta.

Let's check the current relationship in this delta primary. If the line current is 4.33 amps as it reaches the junction point of the two phases it has *two paths* to flow through. Therefore the *phase* current is *less* than the line current. Again the relationship is 1.732 the square root of 3.

Line amps divided by 1.732 = phase amps in a delta (4.33a/1.732 = 2.5a)

Phase amps x 1.732 equal line amps in a delta (2.5a x 1.732 = 4.33a)

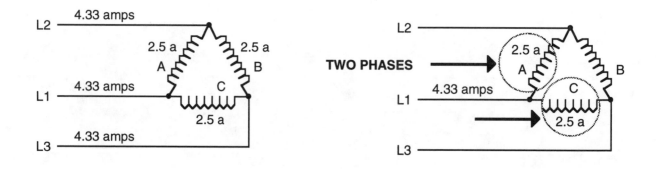

DELTA-WYE
480 / 208 / 120 volt

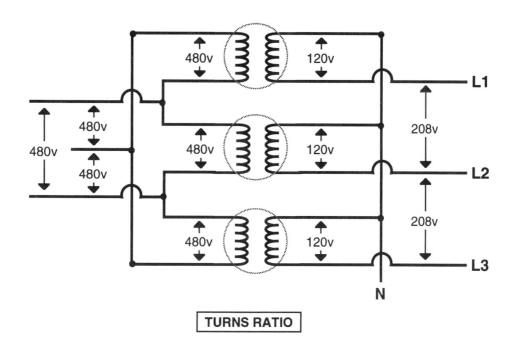

480v

480v

480v

480v
480v
480v

120v
120v
120v

L1

208v

L2

208v

L3

N

TURNS RATIO

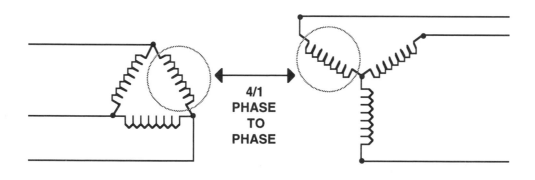

**4/1
PHASE
TO
PHASE**

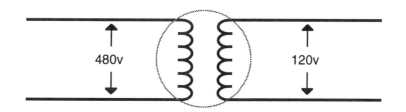

480v

120v

DELTA CONNECTED

PHASE AMPS x 1.732 = LINE AMPS

——— OR ———

$$\frac{\text{LINE AMPS}}{1.732} = \text{PHASE AMPS}$$

PHASE VOLTS = LINE VOLTS

DELTA-DELTA CONNECTED

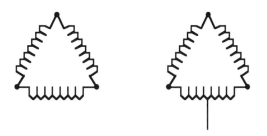

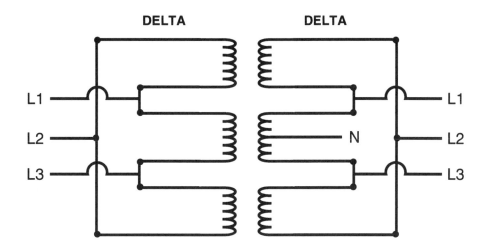

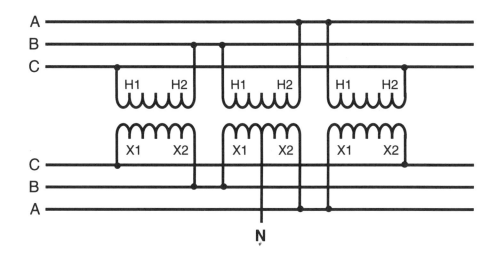

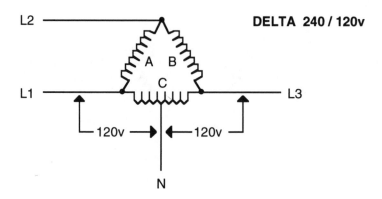

As you can see there is no point in a delta at which an equal potential to all three lines and the grounded neutral can be made.

This is a disadvantage of a delta compared to a wye secondary connection.

The delta secondary connection has only one transformer winding with a neutral conductor, winding "C". The mid-point of winding "C" is tapped which gives the L1 and L3 to neutral a voltage reading of 120 volts.

In a three-phase system transformer "C" is the workhorse, it has to carry all the 120v lighting and appliance loads plus one-third of all the three phase loads.

Transformers "A" and "B" can **not** carry any 120 volt loads as there is no neutral connection to these windings. Transformers "A" and "B" can only carry one-third of the three-phase loads each, and the 240 volt single-phase loads.

The four wire delta system is considered **dangerous** because of the voltage from the high-leg to grounded neutral conductor.

TRANSFORMER "C" will be the largest in a delta 4-wire

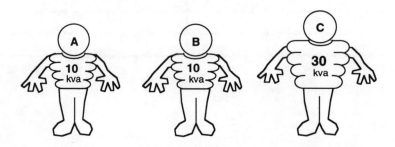

The Code requires the high-leg conductor to be identified by an **orange** color or **tagged**.

To calculate the high-leg to neutral voltage;

Line to neutral voltage x 1.732.

Example: Line to neutral voltage   120 x 1.732 = 208 volts

Calculate the high-leg to neutral voltage on a 230/115v 4-wire delta:

Line to neutral voltage is 115 x 1.732 = **199.18** high-leg to neutral voltage.

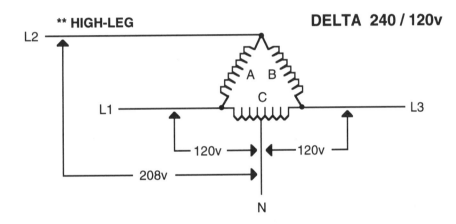

The electrician must use **caution** when making connections in a delta three-phase 4-wire panel **not** to connect any 120 volt loads neutral to L2. The 208 volts would damage the 120 volt equipment.

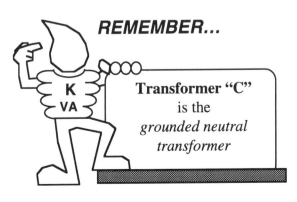

REMEMBER...

Transformer "C" is the *grounded neutral transformer*

THE OPEN DELTA

It is possible to achieve three-phase by using only two transformers. This connection is called the *open delta* or *V-connection*.

Although the open-delta is generally used only as an emergency or temporary system, an original transformer installation may consist of an open-delta bank to supply a three-phase load which is presently light but is expected to increase in the future. This keeps the initial cost low by using only two transformers, a third transformer can be added to the system later when the demand requires it. When the third transformer is added, a delta-delta **closed** bank is formed.

A three-phase transformer with an assembly of three separate single-phase transformers in one tank is lower in initial cost, costs less to install, and requires less space than three separate single-phase transformers.

But, a three-phase transformer has one disadvantage if one of the phase windings becomes defective, the entire three-phase bank must be disconnected and removed from service. A defective single-phase transformer in a three-phase bank can be disconnected and removed for repair. Partial service can be restored using the remaining single phase transformers open-delta until a replacement transformer is obtained. With two transformers three-phase is still obtained, but at reduced power. 57.7% of original power.

This makes it a very practical transformer application for temporary or emergency conditions.

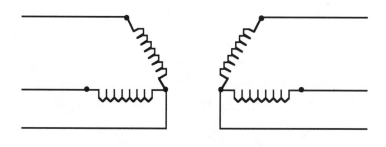

OPEN-DELTA 57.7%

DELTA-DELTA
480 / 240 / 120 volt

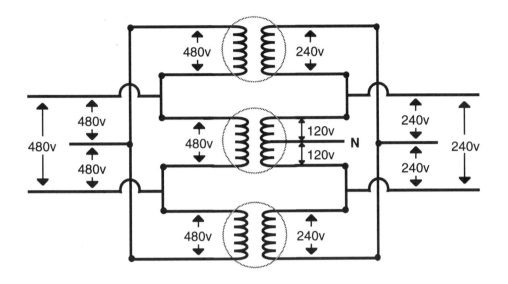

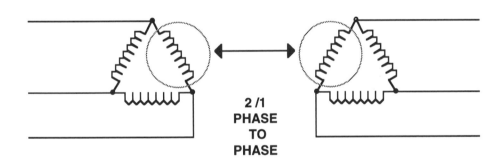

**2 /1
PHASE
TO
PHASE**

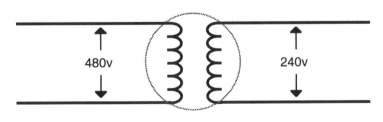

2/1 PHASE TO PHASE

DELTA	WYE
Phase Volts = Line Volts	Phase Volts x 1.732 = Line Volts
Phase Amps x 1.732 = Line Amps	Line Volts / 1.732 = Phase Volts
Line Amps / 1.732 = Phase Amps	Phase Amps = Line Amps

Shown above are the formulas we'll be using to answer transformer exam questions on voltage and current relationships delta or wye connected.

The transformer exam question can be asked in a sentence form or shown in line diagram. If the question is asked in sentence form, sketch out the question into a line diagram. It will make it much easier to calculate.

"A picture is worth a 1000 words".

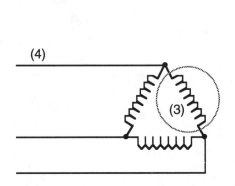

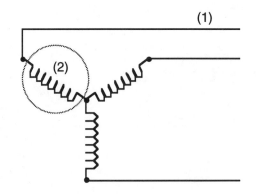

In the sketch we have **four** points to identify:

(1) secondary line   (2) secondary phase   (3) primary phase   (4) primary line

72

EXAM QUESTION EXAMPLE ONE

A delta-wye 480/208/120v three-phase transformer has a secondary line current of 80 amps. What is the primary line current?

(a) 20 amp   (b) 34.6 amp   (c) 50 amp   (d) 80 amp

Solution to question:

First step, change the sentence form question into a sketch:

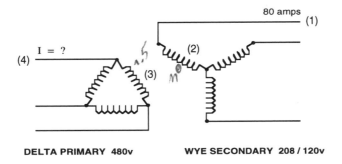

DELTA PRIMARY 480v   **WYE SECONDARY 208 / 120v**

The question stated the (1) secondary line current is 80 amp, the secondary is WYE connected, so using the formula for wye (line amps = phase amps), at point (2) the secondary phase current is also 80 amp. Now from point (2) to point (3) is the ratio of transformation, which in this question 480/208/120v would be phase to phase 480v to 120v a ratio of 4 to 1. At point (2) we have 80 amp, at point (3) we would have 1/4 of the 80 amp as the ratio is 4/1. So the current at point (3) is 80/4 = 20 amp. The primary is DELTA connected, using the formula for delta (phase amps x 1.732 = line amps), 20 amp x 1.732 = 34.64 amp. So the primary line current at point (4) is 34.64 amp the answer to the question.

Now let's check the delta-wye voltage relationship using the formula and checking at points (4), (3), (2), and (1).

The question stated 480/208/120v delta-wye three-phase. This would be a DELTA connected primary at 480 volt, with a WYE connected secondary at 208/120v.

Starting with the primary line voltage at (4) we have 480v, using the formula for delta (line volts = phase volts), we also have 480v in the primary phase at point (3). From point (3) to point (2) is the ratio of transformation, which in this example is phase to phase 480v to 120v a ratio of 4 to 1. So the secondary phase voltage at point (2) is 480/4 = 120 volts. The secondary is WYE connected, using the formula for wye (phase volts x 1.732 = line volts), 120v x 1.732 = 208 volt the secondary line voltage at point (1).

Summary: By making a sketch of the question, then identifying points (1), (2), (3), and (4) and applying the formula for voltage-current relationship for delta or wye connections, it makes the question less difficult to solve.

EXAMPLE TWO

Three 15 kva single-phase transformers are connected delta-wye. The primary line voltage is 480v. What is the primary phase current?

(a) 125 amp   **(b) 54.12 amp**   **(c) 31.25 amp**   **(d) 93.75 amp**

Solution: First step, make a sketch of the question.

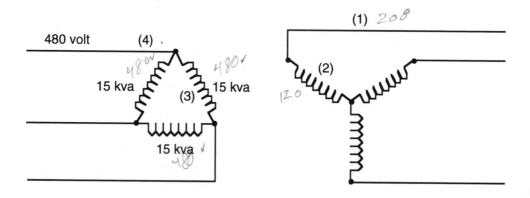

The primary is DELTA connected, using the formula for delta (line volts = phase volts), the phase voltage at point (3) would be 480v same as the line voltage. The phase volt-amps would be 15,000va per phase. To find phase current, use ohms law formula I = W/E = $\dfrac{15000va}{480\ volt}$ = 31.25 amps phase current answer.

To carry this question a step further, the primary line current at point (4) would be phase amps x 1.732 = line amps (from the delta formula), phase amps = 31.25 x 1.732 = 54.12 line amps at point (4).

We know the total power or kva is 3 x 15 kva = 45 kva. Let's check this by using the currents we have found for the phase and line to the question.

Phase current = 31.25 amp x 480 phase voltage = 15,000va per phase, 15,000va x 3 phases = 45,000va total or $\dfrac{45,000va}{1000}$ = 45 kva.

Line current = 54.12 amp x 480 line voltage x 1.732 = $\dfrac{45,000va}{1000}$ = 45 kva.

74

EXAMPLE THREE

A three-phase 2400 / 240v transformer is connected delta-delta. The secondary line voltage is 240, the secondary line current is 150 amps.

1. What is the secondary phase current?
2. What is the primary phase current?
3. What is the primary line current?
4. What is the kva load on the transformer?

Solution: First step make a sketch.

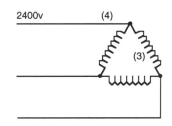 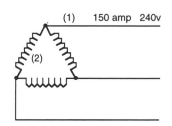

1. To find secondary phase current point (2), the current at point (1) is 150 amp, its DELTA connected secondary, the formula line amps = phase amps,
    $$\frac{\text{line amps}}{1.732} = \text{phase amps},$$
    $$\frac{150 \text{ line amps}}{1.732} = 86.6 \text{ phase amps answer.}$$

2. To find primary phase current point (3), the ratio of transformation occurs between point (2) and point (3), in this example the voltage is $\frac{2400v}{240v}$ 10,

    a ratio of 10 to 1, so the current at point (3) is 86.6 divided by 10 = 8.66 amp.

3. To find the primary line current point (4), the primary is DELTA connected, using the formula for delta (phase amps x 1.732 = line amps), 8.66 x 1.732 = 15 amps primary line current.

4. To find the kva load on the transformer, 240v x 150 amp x 1.732 = $\frac{62,352}{1000}$ = 62.352 kva

    Or using the primary, 2400v x 15 amp x 1.732 = $\frac{62,352va}{1000}$ = 62.352 kva.

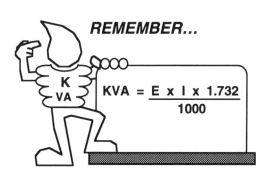

REMEMBER...

$$KVA = \frac{E \times I \times 1.732}{1000}$$

An exam question in diagram form. Delta-wye transformer with a delta load.

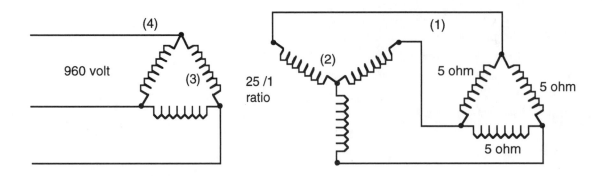

Determine the current in the primary line.

Solution: First step, indicate point numbers (4), (3), (2), and (1) on diagram.

The delta three-phase load is shown as 5 ohm. To change this load to current we first need to know the voltage at this delta load.

Starting at point (4) the delta primary line voltage is 960v. Using the delta formula (line volts = phase volts), we would also have 960 volts at point (3) which is the delta primary phase voltage. From point (3) to point (2) is the ratio of transformation, which is indicated in this calculation as a 25 to 1 ratio. So, the wye secondary phase voltage at point (2) would be 960/25 = 38.4 volts. To find the wye secondary line voltage at point (1), the formula for wye (phase volts x 1.732 = line volts), 38.4 volts x 1.732 = 66.5 volts secondary line.

The 66.5v at point (1) flow directly to the line for the delta load. Using the formula for delta (line volts = phase volts), we have a phase voltage on the delta load of 66.5v. Now we can determine the delta load current using ohms law (I = E / R), 66.5 volts divided by 5 ohm = 13.3 amps current in the delta phase load. To find the current in the delta load line, use delta formula (phase amps x 1.732 = line amps). 13.3 phase amps x 1.732 = 23 amps flowing to point (1).

From point (1) to point (2) is a direct flow from the line to the phase wye connected, so we would also have 23 amps at point (2) the wye secondary phase current.

From point (2) to point (3) is the ratio (25 to 1), 23 amps divided by 25 = .92 amps at point (3) which is the delta primary phase current.

To find delta primary line current, use delta formula (phase amps x 1.732 = line amps), .92 phase amps x 1.732 = 1.59 amps primary line current the answer to the question.

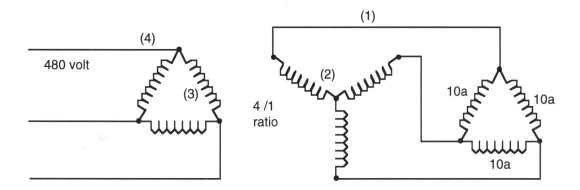

What is the total wattage of the delta connected 10 amp three-phase load?

(a) 2080w **(b) 3602.5w** **(c) 6240w** **(d) 8313.6w**

Solution: First step add point numbers (4), (3), (2), and (1) to diagram.

To find wattage use ohms law formula (W = E x I). Now we need to find the voltage at the 10 amp delta load. Starting at point (4) the delta primary line voltage is 480 volt. Using the formula for delta (line volts = phase volts), we would also have 480 volts in the delta primary phase at point (3).

From point (3) to point (2) is the ratio of transformation (indicated in this calculation as a 4 to 1 ratio). The voltage in the wye secondary phase would be 480 volt divided by 4 = 120 volts at point (2).

To find the wye secondary voltage at point (1), use wye formula (phase volts x 1.732 = line volts), 120 volt x 1.732 = 208 volts at point (1).

With the point (1) voltage 208v flowing directly into the line of the delta load, we need to find the phase voltage. Using the delta formula (line volts = phase volts), we would have the same voltage. So the delta load has a voltage of 208v.

W = E x I   208v x 10 amp = 2080 watts per phase x 3 phases = **6240w answer.**

Or the line current for the delta load would be 10 amp x 1.732 = 17.32 amps

17.32 amps x 208 volt x 1.732 = **6240 watt.**

THREE - PHASE

LOAD

BALANCING

THREE-PHASE LOAD BALANCING

The three-phase 4-wire secondary can be connected only two ways, either WYE or DELTA.

Let's start with the wye 4-wire 208 / 120v three-phase secondary.

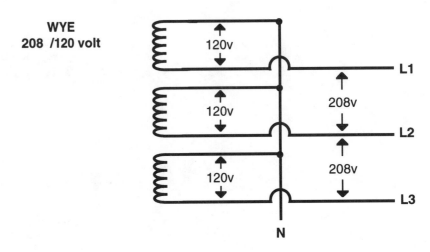

We have three possible voltage connections:

- 208 volt 3ø loads connect   L1 - L2 - L3
- 208 volt 1ø loads connect   L1-L2 or L2-L3 or L3-L1
- 120 volt 1ø loads connect   L1-N L2-N L3-N

Since in either of the three-phase systems, delta or wye, the voltages are **vectors**, the currents derived from loads are **vector quantities**. In order to add these currents, they must be added **vectorially**.

The easiest way to handle transformer loads is to convert all the loads to volt-amps and distribute the volt-amps to the phase to which they are connected.

Division of Loads:

208 volt three-phase loads   - 1/3 on phase A
                              - 1/3 on phase B
                              - 1/3 on phase C

208 volt single-phase loads   - 1/2 on any one phase
                              - 1/2 on any other phase

120 volt single -phase loads - 100% on any one phase

LOAD BALANCING EXAMPLE ONE

Balance the following loads on a 208 / 120v 4-wire three-phase secondary.

1 - 6000va 3ø 208v motor
3 - 5000va 1ø 208v heaters
3 - 150va 1ø 120v lights

Solution: The 6000va three-phase 208v motor divides 1/3 on each phase. The three 5000va heaters would divide 1/2 per phase since the single-phase voltage is 208v, the heaters would connect from L1-L2, L2-L3, and L3-L1. The three 150va lights are single-phase 120 volt, 120v connects from line to neutral. Since we have three 150va loads, we can equally balance them one per phase.

PHASE A	PHASE B	PHASE C
2000va	2000va	2000va
2500va	2500va	2500va
2500va	2500va	2500va
150va	150va	150va
7150va	7150va	7150va

L1 amps	L2 amps	L3 amps
$\dfrac{7150va}{120v} = 59.5$	$\dfrac{7150va}{120v} = 59.5$	$\dfrac{7150va}{120v} = 59.5$

- *Checkpoint: WYE-connected phase amps = line amps,*
  Total va = 7150va x 3 phases = $\dfrac{21,450va}{208v \ x \ 1.732}$ = 59.5 line amps.

NEUTRAL:	PHASE A	PHASE B	PHASE C
	150va	150va	150va

With all loads on, the neutral would carry ZERO current as we have a balance. Maximum neutral current would occur when two 150va loads are OFF, then the neutral would carry $\dfrac{150va}{120v} = 1.25$ amps.

No three-phase loads appear on the neutral, single-phase 208v loads connected across two phases will **not** appear on the neutral. Only 120 volt single-phase loads are connected to the neutral.

Maximum neutral current:	PHASE A	~~PHASE B~~	~~PHASE C~~	
	150va	~~150va~~	~~150va~~	*LOSS OF TWO PHASES*

If any one of the three 150va loads shut off, with two loads on, the neutral will carry the unbalance which would be 150va.

PHASE A	PHASE B	PHASE C	
150va	off - - - - -	150va	*NEUTRAL WOULD CARRY 150va*

80

Example One in a picture form.

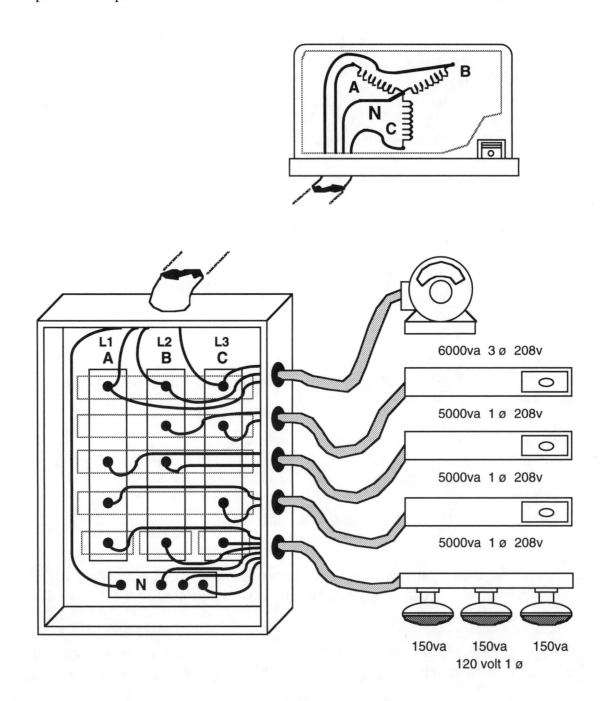

6000va 3 ø 208v

5000va 1 ø 208v

5000va 1 ø 208v

5000va 1 ø 208v

150va   150va   150va
120 volt 1 ø

Delta 4-wire, 240 / 120v three-phase secondary:

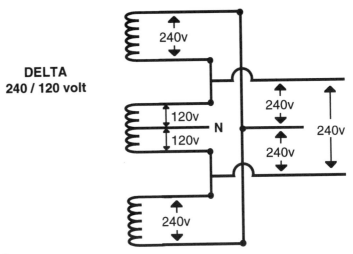

**DELTA
240 / 120 volt**

We have three possible voltage connections:

- 240 volt 3 ø loads connect L1 - L2 - L3
- 240 volt 1 ø loads connect L1-L2 or L2-L3 or L3-L1
- 120 volt 1 ø loads connect L1-N or L3-N

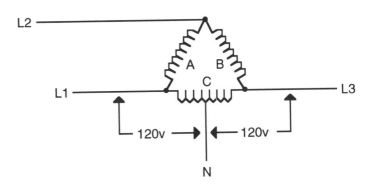

Division of loads:

240 volt 3 ø loads - 1/3 on phase A
                       - 1/3 on phase B
                       - 1/3 on phase C

240 volt 1 ø loads - 100% on any one phase
120 volt 1 ø loads - 100% on phase C only

Let's connect the same loads on a delta 240 / 120v, 4-wire, 3 ø secondary:

1 - 6000va 3 ø 240v motor
3 - 5000va 1 ø 240v heaters
3 -   150va 1 ø 120v lights

Solution: The 6000va three-phase 240v motor divides 1/3 on each phase. The three 5000va heaters 240v single-phase would balance 5000va on each phase. The three 150va lights are single-phase 120 volt so they connect Line One or Line Three to neutral, they can only connect on phase C.

PHASE A	PHASE B	PHASE C
2000va	2000va	2000va
5000va	5000va	5000va
		150va - - - 150va
		150va
7000va	7000va	7450va

$$7000va \ + \ 7000va \ + \ 7450va \ = \ \frac{21{,}450va}{240v \ x \ 1.732} = 51.6 \text{ amps line current}$$

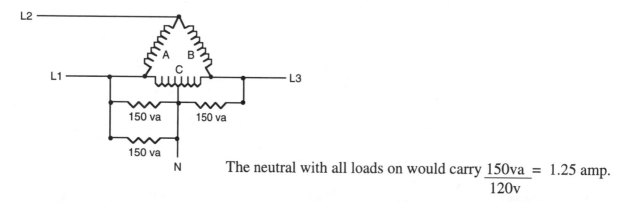

The neutral with all loads on would carry $\frac{150va}{120v} = 1.25$ amp.

Maximum neutral current would occur when the L3 to N load is shut off, then the neutral would have to carry the two L1 to N loads $\frac{300va}{120v} = 2.5$ amps.

By now you can see the differences between a WYE or a DELTA secondary. With a WYE, the 120 volt loads balanced, the 5000va 208v 1 ø loads divided half on each phase. With the DELTA secondary, the 120 volt loads **did not** balance and the 5000va 240v 1 ø loads did balance with one on each phase.

EXAMPLE 2   CONTINUED

Since the delta shows an *unbalance* in the phases, let's calculate the current in each line.

DELTA UNBALANCED

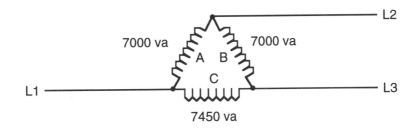

Change to current:

Phase A = 7000va/240v = 29.16 amp
Phase B = 7000va/240v = 29.16 amp
Phase C = 7450va/240v = 31.04 amp

FORMULA FOR UNBALANCED DELTA CURRENT:

L1 current = Phase A + Phase C, divided by 2, times 1.732
L2 current = Phase A + Phase B, divided by 2, times 1.732
L3 current = Phase B + Phase C, divided by 2, times 1.732

Using the above formula calculate the current in the three lines:

L1 = 29.16a + 31.04a = 60.2a/2 = 30.1a x 1.732 = **52.13 amperes in L1**

L2 = 29.16a + 29.16a = 58.32a/2 = 29.16a x 1.732 = **50.5 amperes in L2**

L3 = 29.16a + 31.04a = 60.2a/2 = 30.1a x 1.732 = **52.13 amperes in L3**

EXAMPLE 3

This example shows the difference in current flow in the line conductors using a 45 kva three phases *balanced* and a 45 kva with three phases *unbalanced*.

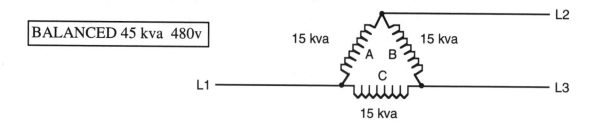

BALANCED 45 kva 480v

15 kva   15 kva

15 kva

L1 current = 15000va/480v = 31.25 phase amps x 1.732 = **54.125 amperes L1**
L2 current = 15000va/480v = 31.25 phase amps x 1.732 = **54.125 amperes L2**
L3 current = 15000va/480v = 31.25 phase amps x 1.732 = **54.125 amperes L3**

In a balanced delta the line current is the *same* in each line 54.125 amperes.

Now let's calculate the same 45 kva total but unbalanced in three phases.

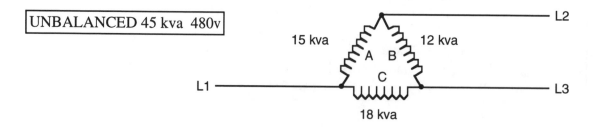

UNBALANCED 45 kva 480v

15 kva   12 kva

18 kva

Phase A current = 15000va/480v = 31.25 phase amps
Phase B current = 12000va/480v = 25 phase amp
Phase C current = 18000va/480v = 37.5 phase amps

L1 current = 31.25a + 37.5a = 68.75a/2 = 34.375a x 1.732 = **59.5 amperes in L1**
L2 current = 31.25a + 25a = 56.25a/2 = 28.125a x 1.732 = **48.7 amperes in L2**
L3 current = 25a + 37.5 = 62.5/2 = 31.25a x 1.732 = **54.1 amperes in L3**

> • **Checkpoint:** L1   59.5a
>   L2 + 48.7a
>   L3 + 54.1a
>   **162.2 amps total divided by 3 = 54.1 per line average**
>   **or**
> $$\frac{45000 \text{ total va}}{480v \times 1.732} = 54.1 \text{ line amps}$$

EXAMPLE 4

DELTA UNBALANCED

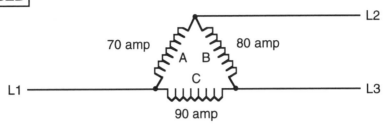

Calculate the current in each line.

L1 current = 70a + 90a = 160a/2 = 80a x 1.732 = **138.6 amperes in L1**

L2 current = 70a + 80a = 150a/2 = 75a x 1.732 = **129.9 amperes in L2**

L3 current = 80a + 90a = 170a/2 = 85a x 1.732 = **147.2 amperes in L3**

• **Checkpoint:** L1 current   138.6
                   L2 current + 129.9
                   L3 current + <u>147.2</u>
                              415.7 divided by 3 = **138.56 ampere average**

Phase A = 70a x 1.732 = 121.24
Phase B = 80a x1.732 + 138.56
Phase C = 90a x 1.732 + <u>155.88</u>
                    415.68 divided by 3 = **138.56 ampere average**

A 10 H.P. 230 volt 3 ø motor has a F.L.C. of 28 amps, three-phase power = E x I x 1.732. 230v x 28a x 1.732 = <u>11,154 total</u> = 3718 per phase.
                     3 phases

A 10 H.P. 208v 3 ø motor has a F.L.C. of 30.8 amps, 208v x 30.8a x 1.732 = <u>11,096 total</u> = 3699 per phase.
                     3 phases

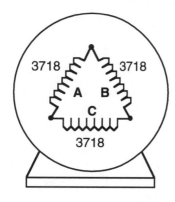
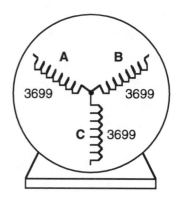

A 5000 watt 1 ø 208v heater connected on a WYE would have 2500w on phase A, and 2500w on phase B.

The same 5000 watt 1 ø heater connected DELTA would have the entire 5000w on phase A.

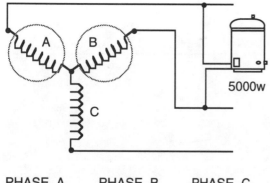
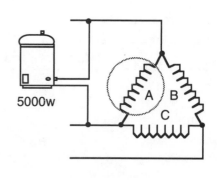

PHASE A   PHASE B   PHASE C   PHASE A   PHASE B   PHASE C
2500w   2500w   --------   5000w   --------   --------

Calculating the neutral current in the WYE secondary.

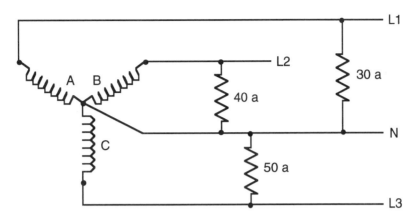

The three loads 30 amp, 40 amp, and 50 amp are connected line to neutral, with all loads "on". What is the unbalanced current flowing in the neutral?

Solution: In $= \sqrt{I2A + I2B + I2C - (IA\ IB) - (IB\ IC) - (IC\ IA)}$

The formula at first looks very difficult, but really it's not.

Everything under the square root sign must be done first, which means:
current in A squared $= 30 \times 30 = 900$
$+$ current in B squared $= 40 \times 40 = 1600$
$+$ current in C squared $= 50 \times 50 = \underline{2500}$
                                 5000 total (call this total "X")

Now the right side of the formula shows:
current in A x current in B $= 30 \times 40 = 1200$
current in B x current in C $= 40 \times 50 = 2000$
current in C x current in A $= 50 \times 30 = \underline{1500}$
                                 4700 (call this total "Y")

Now subtract total "Y" from total "X" $= $ 5000 total X
 — 4700 total Y
 $\overline{300}$

Now extract the square root by pressing the $\sqrt{}$ button on your calculator. The answer is 17.32 amps is the unbalanced current flowing in the neutral. 17.320508 is the square root of 300. 17.320508 x 17.320508 = 300.

This formula is very useful in answering exam questions pertaining to current flow in the neutral of a WYE secondary.

Five 1500 watt 120 volt 1 ø light circuits are connected to a WYE 4-wire secondary 208 / 120v three-phase. The loads are connected 3000 watt, 3000 watt, and 1500 watt per phases A, B, and C. With all loads on, what is the current flowing in the neutral?

(a) 100 amp   (b) 50 amps   (c) 25 amp   (d) 12.5 amp

Solution:	PHASE A	PHASE B	PHASE C
	1500 watt	1500 watt	1500 watt
	1500 watt	1500 watt	- - - - - - -
	3000 watt	3000 watt	1500 watt

The loads are connected 3000-3000-1500. You can see the unbalance would be 1500. If we added 1500 watt to phase C, we would have a balance. But since we have an unbalanced situation, let's check using the formula to see if the correct answer is 1500 watt or 1500 watt = 12.5 amp.

$$\frac{1500 \text{ watt or } 1500 \text{ watt}}{120 \text{ volt}} = 12.5 \text{ amp.}$$

First step convert watts to amps since the question asks for current.

PHASE A

$$\frac{3000 \text{ watt}}{120 \text{ volt}} = 25 \text{ amp}$$

PHASE B

$$\frac{3000 \text{ watt}}{120 \text{ volt}} = 25 \text{ amp}$$

PHASE C

$$\frac{1500 \text{ watt}}{120 \text{ volt}} = 12.5 \text{ amp}$$

Using the formula:

current in A squared	=	25 x 25	= 625
+ current in B squared	=	25 x 25	= 625
+ current in C squared	=	12.5 x 12.5	= 156.25
			1406.25 amp total ("X")

current in A x current in B	=	25 x 25	= 625
current in B x current in C	=	25 x 12.5	= 312.5
current in C x current in A	=	12.5 x 25	= 312.5
			1250 amp total ("Y")

Now subtract total "Y" from total "X" = 1406.25 total "X"
— 1250 total "Y"
156.25 amp

Now extract the square root by pressing the square root button on your calculator. The answer is 12.5 amp. 12.5 amp is the unbalanced current flowing in the neutral.

The maximum neutral current would occur with the loss of either phase A and phase C **or** the loss of phase B and phase C. In either situation, the neutral would have to carry the maximum of 3000 watt or 3000 watt = 25 amp.

$$\frac{3000 \text{ watt or } 3000 \text{ watt}}{120 \text{ volt}} = 25 \text{ amp.}$$

THREE - PHASE

EXAMS

THREE-PHASE EXAM 1

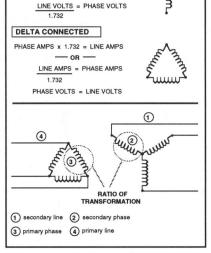

WYE CONNECTED

PHASE AMPS = LINE AMPS

PHASE VOLTS x 1.732 = LINE VOLTS

— OR —

LINE VOLTS = PHASE VOLTS
1.732

DELTA CONNECTED

PHASE AMPS x 1.732 = LINE AMPS

— OR —

LINE AMPS = PHASE AMPS
1.732

PHASE VOLTS = LINE VOLTS

RATIO OF
TRANSFORMATION

① secondary line ② secondary phase
③ primary phase   ④ primary line

1. A delta-wye 480/208/120v three-phase transformer has a secondary line current of 150 amps. What is the primary line current?

(a) 37.5 amps   (b) 65 amps   (c) 150 amps   (d) 3.75 amps

2. The secondary line current of a three-phase 480/208/120v transformer is 95 amp. The kva rating required for this load is approximately _____ kva.

(a) 35   (b) 20   (c) 45   (d) 50

3. A three-phase transformer has a total kva of 12, phase voltage is 120, line current is 33.3 amps. The transformer is connected _____.

(a) delta   (b) wye   (c) cannot tell from the information given

4. What is the turns ratio of a 480/208/120v three-phase transformer?

(a) 480/208   (b) 2/1   (c) 4/1   (d) 1/2

5. The secondary of a three-phase 480/208/120v transformer has a line current of 100 amp. The kva rating of this transformer would be closest to a _____kva transformer.

(a) 20   (b) 40   (c) 60   (d) 50

6. A 480/208/120v 3 ø transformer is connected delta-wye. The secondary line voltage of the transformer is _____ in the secondary.

**(a) less than the phase voltage   (b) equal to the phase voltage
(c) greater than the phase voltage   (d) none of these**

7. Three 10 kva single-phase transformers are connected delta-wye. The primary voltage is 480, what is the primary phase current?

(a) 63 amps   (b) 21 amps   (c) 12 amps   (d) 36 amps

THREE-PHASE EXAM 2

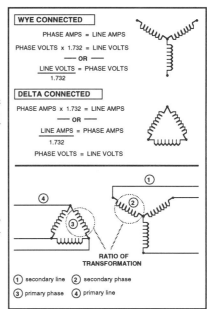

1. If the line current of a three-phase load is 10 amp and the source voltage is three-phase 208/120v, the total 3 ø load in watts would be closest to ____.

(a) 3600   (b) 2080   (c) 1200   (d) none of these

2. If the source is a 75 kva three-phase 480/208/120v transformer, what is the maximum primary line current expected when fully loaded?

(a) 156 amps   (b) 52 amps   (c) 90.2 amps   (d) 30 amps

3. If the secondary line current of a 3 ø delta-wye 480/208/120v transformer is 225 amps, what is the primary phase current?

(a) 56 amps   (b) 97 amps   (c) 130 amps   (d) 390 amps

4. On a 480/240/120v delta transformer, the phase having the higher voltage to the grounded neutral is the high-leg. What is this voltage?

(a) 120v   (b) 190v   (c) 208v   (d) 240v

5. A three-phase delta-wye transformer, 480v primary, 208/120v secondary. The secondary line current is ____.

(a) equal to the secondary phase current   (b) greater than the secondary phase current
(c) less than the secondary phase current   (d) 1.732 x secondary phase current

6. What is the turns ratio on a three-phase 480/240/120v transformer?

(a) 3/1   (b) 2/1   (c) 4/1   (d) 1/2

7. If you were operating three 25 kva single-phase transformers in a delta connected three-phase system and removed one transformer, what is the kva capacity of the two remaining transformers connected delta three-phase?

(a) 28.85 kva   (b) 43.275 kva   (c) 50 kva   (d) 64.95 kva

THREE-PHASE EXAM 3

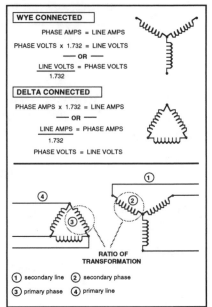

WYE CONNECTED
PHASE AMPS = LINE AMPS
PHASE VOLTS x 1.732 = LINE VOLTS
— OR —
$\frac{\text{LINE VOLTS}}{1.732}$ = PHASE VOLTS

DELTA CONNECTED
PHASE AMPS x 1.732 = LINE AMPS
— OR —
$\frac{\text{LINE AMPS}}{1.732}$ = PHASE AMPS
PHASE VOLTS = LINE VOLTS

RATIO OF TRANSFORMATION

① secondary line   ② secondary phase
③ primary phase   ④ primary line

1. A three-phase 480/240/120v delta-delta transformer has a primary line current of 45 amps. What is the secondary line current?

(a) 22.5 amps   (b) 26 amps   (c) 52 amps   (d) 90 amps

2. In a 3 ø system with a 4-wire, 208/120v, there exists:

(a) 3 ø 208v, 3 ø 120v, and 1 ø 120v
(b) 3 ø 208v, 1 ø 120v, and 1 ø 208v
(c) 3 ø 208v, 3 ø 120v, and 1 ø 208v
(d) 1 ø 208v, 3 ø 208v, and 3 ø 120v

3. The advantage(s) of a 4-wire, 3 ø source over a 3-wire 3 ø source is/are _____.

I. 2 voltage levels   II. a grounded neutral   III. less copper required

(a) I only   (b) II only   (c) III only   (d) I and II

4. If all the load on a three-phase 480/208/120v transformer is single-phase 120v, the maximum neutral current for the 36 kva load would be ____ amps.

(a) 75   (b) 173   (c) 100   (d) 58

5. The secondary of a three-phase 480/208/120v transformer has a line current of 165 amp. The kva rating of this transformer would be closest to a ____ kva transformer.

(a) 35   (b) 40   (c) 50   (d) 60

6. If a wattmeter reads 15,000 watt connected to a three-phase 208/120v load and the power factor measures 0.98, the current would be ____ amps.

(a) 73.59   (b) 41.67   (c) 42.5   (d) 52.0

7. If the total load on a center-tapped 240/120v transformer is 100kw, and the 120v loads were balanced and the 240v load on this phase was 20kw, the maximum neutral current, under any conditions, would be approximately ____ amps.

(a) 0   (b) 192   (c) 333   (d) 667

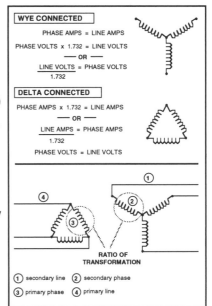

WYE CONNECTED

PHASE AMPS = LINE AMPS

PHASE VOLTS x 1.732 = LINE VOLTS

— OR —

$$\frac{LINE\ VOLTS}{1.732} = PHASE\ VOLTS$$

DELTA CONNECTED

PHASE AMPS x 1.732 = LINE AMPS

— OR —

$$\frac{LINE\ AMPS}{1.732} = PHASE\ AMPS$$

PHASE VOLTS = LINE VOLTS

RATIO OF
TRANSFORMATION

① secondary line ② secondary phase
③ primary phase ④ primary line

1. The primary of a single-phase transformer draws 6 amps at 7200 volts, the secondary shows a reading of 165 amps at 240 volts. What is the efficiency of this transformer?

(a) 21.6%   (b) 58%   (c) 72%   (d) 92%

2. If the secondary line current of a three-phase delta-wye 480/208/120v transformer is 195 amp, what is the primary line current?

(a) 28 amp   (b) 84.4 amp   (c) 48.75 amp   (d) 195 amp

3. On a 460/230/115v delta-delta transformer with a three-phase 4-wire secondary, the phase having the higher voltage to the grounded neutral is the high-leg. What is this voltage?

(a) 208v   (b) 230v   (c) 240v   (d) 199v

4. A three-phase 4-wire wye connected transformer has a current of 25 amps on L1, 20 amps on L2, and 15 amps on L3. These loads are line to neutral. What is the current flowing in the neutral?

(a) 8.66 amp   (b) 20 amp   (c) 30 amp   (d) 50 amp

5. The system is 240/120v three-phase high-leg, the load consists of:

10 - 20 amp branch-circuit lighting loads with each circuit loaded 80%
 1 - 6 kw 240v 1 ø load
 1 - 9 kw 240v 3 ø load

If only the three-phase load is attached to the high-leg and all other loads are supplied by the neutral transformer, what is the load on the grounded neutral transformer?

(a) 19.2 kw   (b) 25.2 kw   (c) 28.2 kw   (d) 34.2 kw

6. If the line current of a three-phase load is 20 amp and the source voltage is three-phase 208/120v, the total 3 ø load in watts would be closest to _____.

(a) 4160   (b) 12480   (c) 7200   (d) 2400

THREE-PHASE EXAM 5

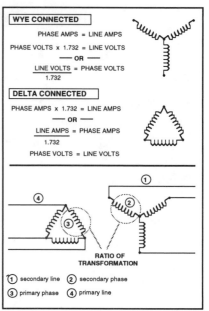

WYE CONNECTED

PHASE AMPS = LINE AMPS

PHASE VOLTS x 1.732 = LINE VOLTS

— OR —

$$\frac{LINE\ VOLTS}{1.732} = PHASE\ VOLTS$$

DELTA CONNECTED

PHASE AMPS x 1.732 = LINE AMPS

— OR —

$$\frac{LINE\ AMPS}{1.732} = PHASE\ AMPS$$

PHASE VOLTS = LINE VOLTS

RATIO OF
TRANSFORMATION

① secondary line ② secondary phase
③ primary phase ④ primary line

1. A three-phase 460/230/115v delta-delta transformer has a primary line current of 62 amps. What is the secondary line current?

(a) 62 amp   (b) 35.79 amp   (c) 71.59 amp   (d) 124 amp

2. Five - 2kw 120v 1 ø light banks are connected on a 4-wire three-phase 208/120v system. Each of the five light banks has a phase load of _____ amps.

(a) 2000   (b) 83   (c) 16.6   (d) 50

3. A 4-wire three-phase 480/277v secondary has three-15kw 277v lighting loads with a 0.80 power factor. What is the line current?

(a) 20-30 amp   (b) 40-50 amp   (c) 60-70 amp   (d) 80-100 amp

4. If the source is a 25 kva three-phase 480/208/120v transformer, what is the maximum primary line current expected when fully loaded?

(a) 17 amp   (b) 30 amp   (c) 52 amp   (d) none of these

5. A three-phase delta-delta high-leg transformer has a primary of 480 volts and a secondary of 240/120v. One of the loads consists of 240 kva 1 ø 120v balanced, the maximum current in the primary of the neutral transformer due to this load would be _____ amperes.

(a) 1000   (b) 2000   (c) 500   (d) none of these

6. Four - 3000 watt 208v single-phase heat strips, they are connected to a 208/120v 3 ø 4-wire secondary. Each of the four strips has a phase load of _____ watts.

(a) 3000   (b) 1500   (c) 12,000   (d) 15,000

7. What is the primary line current when a 15 kva three-phase 480/208/120v transformer is fully loaded at a power factor of 80%.

(a) 22.5 amp   (b) 32 amp   (c) 40 amp   (d) 53.6 amp

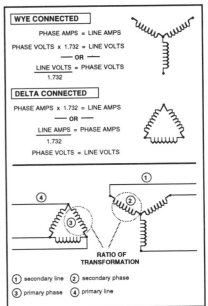

1. Three - 2 hp single-phase 230v motors are balanced on a three-phase 230/115v secondary. What would be the secondary line current if all the motors are running fully loaded?

(a) **12 amp** (b) **21 amp** (c) **27 amp** (d) **30 amp**

2. A 4-wire three-phase 480/277v secondary, what is the voltage line to neutral?

(a) **120v** (b) **208v** (c) **277v** (d) **480v**

3. Three - 4500 watt 208v single-phase dryers are connected to a 208/120v 4-wire secondary. Each of the dryers has a line to line load of ____ watts.

(a) **2250** (b) **4500** (c) **13,500** (d) **6750**

4. Three 15 kva single-phase transformers are connected delta-wye. The primary voltage is 480. What is the primary phase current?

(a) **31 amp** (b) **54 amp** (c) **94 amp** (d) **none of these**

5. What current does a 25 kw load draw from a 240v three-phase line when the power factor is 82%?

(a) **24 amp** (b) **60 amp** (c) **73 amp** (d) **127 amp**

6. Three - 5000 watt 240v single-phase water heaters are connected to a delta 240v three-phase secondary. Each heater has a phase load of ____ watts.

(a) **2500** (b) **5000** (c) **10,000** (d) **15,000**

7.

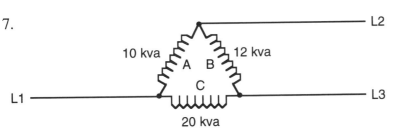

What is the current in L1 of this unbalanced 480v 3 ø delta transformer?

(a) **54 amp** (b) **62 amp** (c) **125 amp** (d) **151 amp**

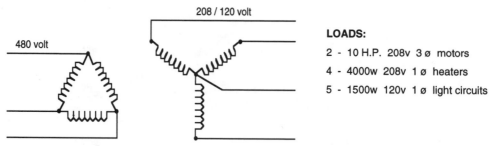

THREE-PHASE EXAM 7

480 volt

208 / 120 volt

LOADS:

2 - 10 H.P. 208v 3 ø motors

4 - 4000w 208v 1 ø heaters

5 - 1500w 120v 1 ø light circuits

1. What is the turns ratio of this transformer?

(a) 2/1   (b) 480/208   (c) 4/1   (d) 208/480

2. If the light circuits are connected 3000w, 3000w, and 1500w per phase A-B-C, what is the maximum neutral load?

(a) 4500w   (b) 3000w   (c) 1500w   (d) 7500w

3. What is the maximum load per secondary conductor due to the three-phase motors?

(a) 28 amp   (b) 30.8 amp   (c) 56 amp   (d) 61.6 amp

4. If the loads are balanced as close as possible with the 120v light circuits connected as indicated in question #2 on the three transformers, what is the minimum kva transformer for each 1 ø transformer if rounded off to the next higher 5 kva?

(a) 20-20-20   (b) 15-15-20   (c) 25-30-30   (d) 15-15-15

5. What is the current in the primary conductors with all loads on?

(a) 127 amp   (b) 35 amp   (c) 55 amp   (d) 95 amp

6. If the transformer is 80% efficient, how does this affect the answer to question #5?

(a) increases 20%   (b) decreases 20%   (c) increases 25%   (d) decreases 25%

7. If all loads are connected as mentioned in question #2 and operating, what is the wattage on the neutral conductor?

(a) 1500w   (b) 3000w   (c) 4500w   (d) 7500w

8. What size 3 ø transformer is required for these loads?

(a) 30 kva   (b) 45 kva   (c) 60 kva   (d) 75 kva

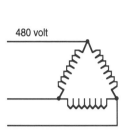

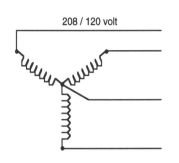

208 / 120 volt

480 volt

LOADS:

1 - 15 H.P. 3 ø 208v motor

4 - 3000w light banks 120v

5 - 2500w heat strips 208v 1 ø

2 - 1500w dryer 208v 1 ø

1 - 800w cook unit 208v 1 ø

1. The total load for the three-phase motor is closest to _____ watts.

(a) 9610   (b) 11,000   (c) 15,000   (d) 16,644

2. Each of the four light banks is connected _____ watts phase to neutral.

(a) 1000   (b) 1300   (c) 3000   (d) 12,000

3. The load per phase of the motor is closest to _____ watts.

(a) 3203   (b) 4117   (c) 5548   (d) 8322

4. Each of the two 1500 watt dryers has a line to line load of _____ watts.

(a) 750   (b) 1500   (c) 3000   (d) 2250

5. If the largest phase load is 15,198w, the secondary line current is ___ amps.

(a) 42   (b) 63   (c) 73   (d) 127

6. With the 120v light banks connected 6000w, 3000w, and 3000w per phases A-B-C, what is the maximum neutral current in amps?

(a) 25   (b) 50   (c) 75   (d) 100

7. Each of the five heat strips has a phase load of _____ watts.

(a) 1250   (b) 2500   (c) 12,500   (d) 3750

8. With a total load of 44,944w on the 3 ø system, what is the primary phase current?

(a) 54 amp   (b) 31 amp   (c) 94 amp   (d) 124 amp

THREE-PHASE EXAM 9

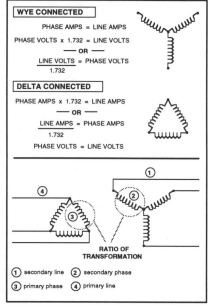

1. If a three-phase, delta-wye transformer bank having a 480v primary and a 208/120v secondary, is considered to be 100% efficient, and to have a resistive-type load, the maximum kva of the load will be ____.

I. equal to the kva of the secondary of the transformer
II. equal to the primary kva of the transformer
III. considerably less than the kva of the transformer

(a) I only   (b) II only   (c) I and II only   (d) I, II and III

2. If a three-phase wattmeter reads 12,442 watts on a three-phase 208/120v system, and each ammeter reads 36 amps, the power factor would be ____.

(a) unity   (b) 0.96   (c) 0.55   (d) none of these

3. Three - 2 hp single-phase 208v motors are balanced on a 208/120v three-phase secondary. The current on each phase is closest to ____ amps.

(a) 12   (b) 18   (c) 24   (d) 26

4. Delta-wye connected transformer 480v primary, 208/120v secondary. The primary line current is ____.

I. phase amps x 1.732
II. equal to phase amps
III. equal to kva divided by 1.732 times 480v

(a) I only   (b) II only   (c) III only   (d) I and III

5. The maximum load on a three-phase flourescent lighting panel is 96 kw. The maximum load on the neutral would be ____ kw.

(a) 96   (b) 67.2   (c) 32   (d) 22.4

6. Three-phase delta-wye , 480v primary, 208/120v secondary, the secondary line current is ____.

(a) equal to the secondary phase current   (b) greater than the secondary phase current
(c) less than the secondary phase current   (d) 1.732 times the secondary phase current

99

THREE-PHASE EXAM 10

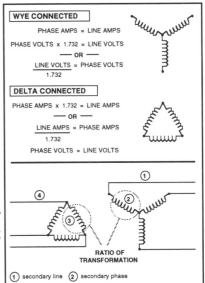

1. To handle a three-phase unbalanced load, balance the system by making all loads equal to the _____ single phase load.

(a) smallest (b) average (c) largest (d) unbalanced

2. If the source is a 50 kva three-phase 480/208/120v transformer delta-wye, what is the maximum primary line current expected at full load?

(a) 35 amp (b) 60 amp (c) 104 amp (d) 123 amp

3. If the secondary load of a 3 ø 480/208/120v transformer is:

9 - 2000 watts of 120v lighting
9 - 2000 watts of 208v 1 ø loads

The neutral must have a load current rating of at least _____ amps.

(a) 50 (b) 75 (c) 86.5 (d) 150

4. Three - 10 kva single-phase transformers are connected delta-wye. The primary voltage is 480v. What is the maximum primary line current?

(a) 62.5 amps (b) 36 amps (c) 20.8 amps (d) 12 amps

5. Given: Δ load, V phase = 208, Z phase = 10Ω

I line = _____ amps.

(a) 20.8 (b) 12 (c) 36 (d) none of these

6.

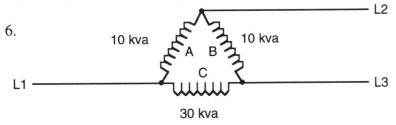

Three-phase 240v transformer, what is the current in L3?

(a) 50 amp (b) 120 amp (c) 208 amp (d) 144 amp

100

ANSWERS

SINGLE-PHASE EXAM 1 ANSWERS

1. **(c) 5/1** To find the turns ratio of this transformer divide the higher voltage by the lower voltage 120v/24v = 5 or 5 to 1 ratio.

2. **(a) 10 amp** To find the secondary current use ohms law formula for current I = W/E 240w/24v = 10 amp.

3. **(c) 2 amp** To find primary current use ohms law formula I = W/E. Remember at 100% efficiency the watts in the primary are the same as the watts in the secondary, 240w. The difference would be the primary voltage is 120v. 240w/120v = 2 amp primary current.
 • Checkpoint: The turns ratio is 5 to 1, so the secondary would have 5 times the current that the primary has. Primary current = 2 amp x 5 = 10 amp secondary current.

4. **(d) 240va** To find the volt-ampere input on the primary, the va in the primary is the same as the va in the secondary at 100% efficiency.

SINGLE-PHASE EXAM 2 ANSWERS

1. **(a) 300w** To find the load add 100w + 100w + 100w = 300w load

2. **(c) 2.5 amps** To find the current in the primary use the ohms law formula for current I = W/E 300w/120v = 2.5 amperes primary current.

3. **(c) 10/1 ratio** To find turns-ratio divide the higher voltage by the lower voltage 120v/12v = 10.

4. **(b) 25 amps** To find the secondary current use ohms law formula for current I = W/E 300w/12v = 25 amperes secondary current

SINGLE-PHASE EXAM 3 ANSWERS

1. **(c) 5/1 ratio** To find turns-ratio divide the higher voltage by the lower voltage 120v/24v = 5.

2. **(c) 20.83 amps** To find secondary current use ohms law formula I = W/E = 500w/24v = 20.83a.

3. **(b) 4.16 amps** To find the primary current I = W/E = 500va/120v = 4.16 current in the primary.

4. **(a) 500va** To find the input va at 100% efficiency it is equal to the output va which is 500.

SINGLE-PHASE EXAM 4 ANSWERS

1. **(b) 20.8 amps** To find the secondary current I = W/E = 500va/24v = 20.8 amperes.

2. **(b) 521va** To find the input va INPUT = OUTPUT/EFFICIENCY = 500va/.96 = 521va input.
•By the efficiency going down to 96% the input va went up to 521 for the output to remain the same at 500va.

3. **(c) 4.34 amps** To find the primary current I = W/E = 521va/120v = 4.34 amps primary current.

4. **(d) .521 kva** To find the primary kva divide the va by 1000, 521va/1000 = .521 kva.
• K =1000   521va = .521 kva   va = volt-amps   kva = thousand-volt-amps

1. **(b) 82%**   PF = W/VA = 6500w/240v x 33a = 82%

2. **(a) 86%**   PF = W/VA = 2000w/230v x 10.1a = 86%

3. **(c) 74.4 amps**   I = W/E x PF = 15,000w/240v x .84 = 74.4 amps

4. **(b) 8500w**   W = VA x PF = 10,000va x .85 = 8500w

5. **(b) 86%**   PF = OUTPUT/INPUT = 21,500w/25,000va = 86%

6. **(d) 235 volts**   E = W/I x PF = 5000w/25a x .85 = 235 volts

7. **(b) .905**   EFF = OUTPUT/INPUT = 20,000w/92a x 240v = .905

8. **(b) 72.67 amps**   I = W/E x PF = 15,000w/240v x .86 = 72.67 amps

1. **(d) 8/1 ratio**   Ratio = higher voltage/lower voltage = 120v/15v = 8 ratio

2. **(b) 4.16 amps**   I = W/E = 100w/24v = 4.16 amperes

3. **(a) 3660va**   INPUT = OUTPUT/EFF = 3550va/.97 = 3659.7 or 3660va

4. **(a) 2 amps**   I = W/E = 240va/120v = 2 amperes

5. **(b) secondary**   Secondary carries more current, higher the current the larger the wire

6. **(a) 10 amps**   The neutral carries the unbalanced current = 1200w/120v = 10 amps

7. **(a) 86%**   PF = W/VA = 7500w/230v x 38a = 86%

8. **(b) 2.5 amps**   I = W/E = 300w/120v = 2.5 amperes

1. **(c) 2/1 ratio**   Ratio = 480v/240v = 2   •Remember 120v is **not** phase voltage, it's half a phase

2. **(c) 59.28 amps**   I = W/E x PF = 12,000w/230v x .88 = 59.28 amperes

3. **(d) zero**   In a balanced circuit the neutral would carry zero current

4. **(c) 2.17 amps**   INPUT = OUTPUT/EFF = 500va/.96 = 521va input = 521va/240v = 2.17 amps

5. **(b) 93%**   EFF = OUTPUT/INPUT = 33,600w/36,000va = 93%

6. **(c) 3.67 kva**   120v x 30a = 3600va = 3600va/.98 = 3673va/1000 = 3.67 kva

7. **(a) 35 amps**   L1 = 1500w + 1500w + 1200w = 4200w
                      L2 = 1500w + 1500w   = 3000w
The maximum neutral would be L1 load 4200w/120v = 35 amperes

SINGLE-PHASE EXAM 8 ANSWERS

1. **(c) 30/1** Turns-ratio = 7200v/240v = 30 •Remember, 120v is only **half** the phase voltage.

2. **(c) 76.75 amps** To find the current in L1, change all loads to watts.
$W = E^2/R = 120v \times 120v/ 15\Omega = 960w$ $W = E \times I = 120v \times 10a = 1200w$
L1 will carry half of the 9600w = 4800w   L2 will carry half of the 4.5kw = 2250w
Add the four L1 loads, 960w + 1200w + 4800w + 2250w = 9210 total watts L1
$I = W/E = 9210w/120v = 76.75$ amperes L1 current

3. **(a) 85.75 amps** To find current in L2, change all the loads to watts.
$W = E \times I = 12a \times 120v = 1440w$ $W = E^2/R = 120v \times 120v/8\Omega = 1800w$
L2 will carry half of the 9600w = 4800w   L2 will carry half of the 4.5kw = 2250w
Add the four L2 loads, 1440w + 1800w +4800w + 2250w = 10,290 total watts L2
$I = W/E = 10,290w/120v = 85.75$ amperes L2 current

4. **(d) 9 amps** To find the neutral current, the unbalanced on the neutral would be the L2 neutral load of 27 amps minus the L1 neutral load of 18 amps = 9 amps neutral current. •Remember, the 9600w and the 4.5kw are 240v connected and are **not** neutral loads.

5. **(d) 27 amp** The maximum neutral current would be the L1 loads shut off, the neutral would have to carry the L2 to neutral load of 27 amperes.

6. **(d) 19.5 kva** To find kva, add L1 power 9210w + L2 power 10,290w = 19,500w/1000 = 19.5

7. **(d) 81.25 amps** To find secondary current, I = W/E = 19,500w/240v = 81.25 amperes

8. **(a) 2.71 amps** To find primary current, I = W/E = 19,500w/7200v = 2.71 amperes
•Remember, 30/1 ratio, secondary current is 81.25a/30 ratio = 2.71 amps

1. **(d) 3242va**   VA = W/PF = 3080W/.95 = 3242va

2. **(a) .94**   PF = W/VA = 2900w/3080va = .94

3. **(c) .5 kva**   500w in secondary = 500w in primary @ 100% EFF   500/1000 = .5 kva

4. **(b) 45 kva**   va in primary = va in secondary @ 100% EFF   45 kva

5. **(c) is true**

6. **(b) 104 amp**   50 kva x 1000 = 50,000va/480v = 104 amp

7. **(b) 100 amps**   48 kva - 24 kva = 24 kva balanced on neutral, 12kv L1 to neutral and 12 kva L2 to neutral. Maximum unbalance = 12 kva. 12,000va/120v = 100 amperes.

SINGLE-PHASE EXAM 10 ANSWERS

1. **(c) 2/1 ratio**   480v/240v = 2 to 1 ratio •Remember 120v is **not** phase voltage, it is one-half.

2. **(d) 92%**   EFF = OUTPUT/INPUT = 39,600w/43,200va = .916 or 92%

3. **(b) 16 amps**   2 hp 1 ø 120v F.L.C. = 24a x 120v = 2880va
L1 = 2880 + 1900 + 1900 = 6680   L2 = 2880 + 1900 = 4780
unbalance of 1900/120v = 15.8 or 16 amps •Remember, if one motor is on each hot wire.

4. **(c) is false**

5. **(d) 2.94 kva**   24a x 120v = 2880va/.98 = 2938.7va/1000 = 2.9387 or 2.94 kva

6. **(d) 75 amps**   The maximum would be 9000w/120v = 75 amperes

SINGLE-PHASE EXAM 11 ANSWERS

1. **(b) .83**   PF = W/VA = 600w/720va = .83

2. **(c) is true**

3. **(b) is false**

4. **(d) 240 volts**   32v/40 turns = .8 volt per turn   .8v per turn x 300 turns = 240v

5. **(a) 127 amp**   I = W/E x PF = 25,000w/240v x .82 = 127 amperes

6. **(d) 1 amp**   I = W/E = 120w/120v = 1 ampere

SINGLE-PHASE EXAM 12 ANSWERS

1. **(b) 5000 watts**   OUTPUT = INPUT x EFFICIENCY   5263va x .95 = 4999.85 or 5000 watts

2. **(c) .5 amp**   2 hp 1 ø 120v motor F.L.C. = 24a x 120v = 2880va
L1 = 1900 + 1900 + 1900 = 5700   L2 = 2880 + 2880 = 5760
unbalance difference of 60va/120v = .5 ampere

3. **(a) 10 amp**   With 9 - 1200w 120v loads L1 would have 5 loads to neutral and L2 would have 4 loads to neutral, the unbalance would be one 1200w load/120v = 10 ampere. •Remember the water heater at 5000w and the heater at 6000w are 240v connected, they do **not** connect to the neutral.

4. **(b) 104 amp**   I = W/E = 25,000va/240v = 104 amperes

5. **(d) 240 volt**   E = W/I x PF = 10,000w/50a x .835 = 239.5 or 240 volts

6. **(c) 1.25 amperes**   I = W/E = 150w/120v = 1.25 amperes

1. **(b) 65 amps**   150a/4 ratio = 37.5a primary phase x 1.732 = 65 primary line amps

2. **(a) 35 kva**   3 ø KVA = E x I x 1.732/1000   208v x 95a x 1.732 = 34,224va/1000 = 34.2 or 35

3. **(b) wye**   12,000va total/3 phases = 4000va per phase/120v = 33.3 phase amps •Remember, phase amps = line amps in a wye, 33.3 phase amps, 33.3 line amps. Wye connected.

4. **(c) 4/1 ratio**   Ratio = 480v/120v = 4 to 1. **Use phase to phase voltage.** 208 is a **line** voltage.

5. **(b) 40 kva**   208v x 100a x 1.732 = 36,025va/1000 = 36 kva or 40 kva

6. **(c)**   208 line voltage is greater than the 120 phase voltage

7. **(b) 21 amps**   10,000 va per phase/480 phase volts = 20.8 or 21 phase amps

1. **(a) 3600 watts**   3 ø power = E x I x 1.732   208v x 10 amp x 1.732 = 3602 or 3600w total

2. **(c) 90.2 amps**   I = W/E x 1.732 = 75,000va/480v x 1.732 = 90.2 amps primary line

3. **(a) 56 amps**   225a/4 ratio = 56.25 amps primary phase

4. **(c) 208v**   120v x 1.732 = 208v high-leg to neutral voltage

5. **(a)**   In a wye the line current is equal to the phase current

6. **(b) 2/1 ratio**   480v/240v = 2 to 1 ratio •Remember, 120v is half phase voltage, **not** phase voltage

7. **(b) 43.275 kva**   75,000va x 57.7% = 43,275va/1000 = 43.275kva

THREE-PHASE EXAM 3 ANSWERS

1. **(b) 26 amps**   45a/1.732 = 26 amps primary phase

2. **(b)**   3 ø 208v, 1 ø 120v, and 1 ø 208v

3. **(d)**   2 voltage levels and a grounded neutral

4. **(c) 100 amps**   PHASE A   PHASE B   PHASE C   12,000va/120v = 100 amperes
   12000va   ~~12000va~~   ~~12000va~~

5. **(d) 60 kva**   208v x 165a x 1.732 = 59,442va/1000 = 59.442 kva or 60 kva

6. **(c) 42.5 amps**   I = W/E x PF x 1.732   15,000w/208v x .98 x 1.732 = 42.5 amperes

7. **(c) 333 amps**   Total 100kw, 20kw is 240v connected, this leaves a total of 80kw left which is 120v balanced, balanced would mean that you have 40kw L1 to neutral and 40kw L2 to neutral. The maximum unbalanced would occur if either line was shut off, leaving a maximum neutral load of 40kw, 40,000w/120v = 333 amperes.

1. **(d) 92%** EFF = OUTPUT/INPUT 165a x 240v = 39600w output 6a x 7200v = 43200va input
39600w/43200va = .916 or 92%

2. **(b) 84.4 amps**

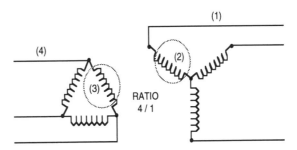

① = 195 amp

② = 195 amp

③ = 195a/4 ratio = 48.75a

④ = 48.75a x 1.732 = 84.4 amperes

3. **(d) 199 volt** 115 volt x 1.732 = 199.18 volts high-leg to neutral voltage

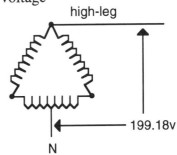

4. **(a) 8.66 amp**

25a x 25a = 625a	25a x 20a = 500a	1250a
20a x 20a = 400a	20a x 15a = 300a	-1175a
15a x 15a = 225a	15a x 25a =375a	75a
1250a	1175a	

$\sqrt{75a}$ = 8.66a

5. **(c) 28.2 kw**

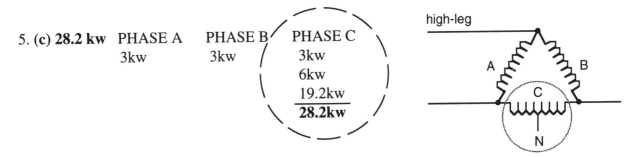

PHASE A	PHASE B	PHASE C
3kw	3kw	3kw
		6kw
		19.2kw
		28.2kw

lighting circuit 10 x 20a x 120v x 80% = 19.2kw. The **grounded neutral transformer** is "C".

6. **(c) 7200w** 20a x 208v x 1.732 = 7205w

1. **(d) 124 amp**

   ④ = 62a

   ③ = 62a/1.732 = 35.79a

   ② = 35.79a x 2 ratio = 71.59a

   ① = 71.59a x 1.732 = 124 amperes

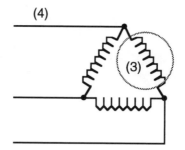
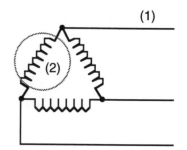

2. **(c) 16.6 amp** *EACH* has a 2000w phase to neutral load. 2000w/120v = 16.6 amperes

3. **(c) 60-70 amps** Phase load = 15,000w/277v x .80 = 67.6 phase amps
Wye connected phase amps = line amps. 45,000w/480v x .80 x 1.732 = 67.6 line amps

4. **(b) 30 amp** 25,000va/480v x 1.732 = 30 amp

5. **(c) 500 amperes** Maximum neutral = 120,000va/120v = 1000a in secondary
ratio 2/1, primary neutral transformer "C" would have to carry 1000a/2 = 500 amperes

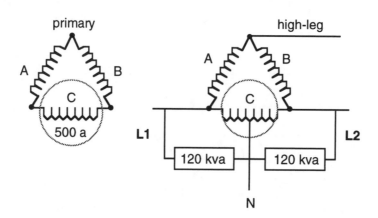

6. **(b) 1500 watts** *EACH* has a phase load of
3000w/2 phases = 1500 watt per phase

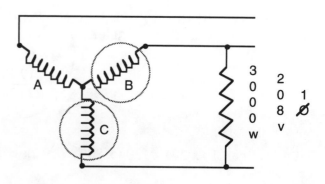

7. **(a) 22.5 amp** 15000va/480v x .80 x 1.732 = 22.5 amperes

1. **(b) 21 amp**   2 hp 1 ø 230v F.L.C. = 12a   Delta secondary phase amps 12 x 1.732 = 20.7 or 21 line amps.

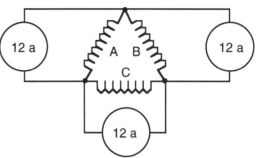

2. **(c) 277 volt**   480v/1.732 = 277v

3. **(b) 4500 watt**   208v 1 ø 4500w **EACH** line to line

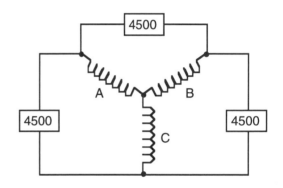

4. **(a) 31 amp**   15,000va/480v = 31 phase amps

5. **(c) 73 amp**   25,000w/240v x .82 x 1.732 = 73 amperes

6. **(b) 5000 watts**   Delta secondary 240v 1 ø **EACH** has a **PHASE** load of 5000 watts.

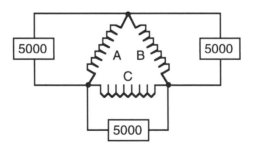

7. **(a) 54 amp**   Phase A current = 10,000va/480v = 20.83a   Phase C current = 20,000va/480v = 41.66a
20.83a + 41.66a = 62.49a/2 = 31.245a x 1.732 = 54 amperes

1. **(c) 4/1 ratio**   480/120v = 4 to 1 phase to phase voltage transformation, 208v is a **line** voltage.

2. **(b) 3000 watts**   The maximum neutral load would be 3000w with the loss of the 1500w load and either 3000w load.

3. **(d) 61.6 amps**   10 hp 208v 3 ø motor F.L.C. = 30.8a x 2 motors = 61.6 amperes

4. **(b) 15-15-20**

PHASE A	PHASE B	PHASE C
3699w	3699w	3699w
3699w	3699w	3699w
2000w	2000w	------
------	2000w	2000w
2000w	------	2000w
2000w	------	2000w
1500w	1500w	1500w
1500w	1500w	------
16,398w	14,398w	14,898w

USE 20 kva   USE 15 kva   USE 15 kva

5. **(c) 55 amp**   45,694w/480v x 1.732 = 54.96 or 55 amperes

6. **(c) increases 25%**   INPUT = OUTPUT/EFF = 45,694w/.80 = 57,117.5w INPUT
To find percent: 57,117.5w/45,694w = 1.25 or a 25% **increase**.

7. **(a) 1500 watt the unbalanced**

PHASE A	PHASE B	PHASE C
1500w	1500w	1500w
1500w	-------	1500w
3000w	1500w	3000w

8. **(c) 60 kva 3 ø transformer**   3 x largest single-phase transformer = 3 x 20kva = 60 kva.

1. **(d) 16,644**   F.L.C. = 46.2a x 208v x 1.732 = 16,644

2. **(c) 3000**   The key word is **EACH**, each light bank is 3000 to neutral

3. **(c) 5548**   Total is 16,644/3 phases = 5548w per phase

4. **(b) 1500**   The key is **LINE to LINE**, line to line is 208v, 1500 each

5. **(d) 127 amp**   15,198w/120v = 127 phase amps.   **WYE PHASE AMPS = LINE AMPS**

PHASE A	PHASE B	PHASE C
5548w	5548w	5548w
3000w	3000w	3000w
3000w	-------	-------
1250w	1250w	-------
1250w	-------	1250w
-------	1250w	1250w
-------	1250w	1250w
-------	1250w	1250w
-------	750w	750w
750w	-------	750w
400w	400w	-------
15,198w	14,698w	15,048w

6. **(b) 50 amp**

PHASE A	PHASE B	PHASE C
3000w	3000w	3000w
3000w	-------	--------
6000w	3000w	3000w

With the loads on phases B and C off, the maximum neutral current would be 6000w/120v = 50a

7. **(a) 1250**   The key word is **EACH**, and **PHASE LOAD**. Remember a 208v 1 ø load connects across two phases on a wye to get the 208v. So each **phase** carrys one-half the load or 2500w/2 = 1250w per phase.

8. **(b) 31 amp**   The key word is **PHASE** current.

44,944w/480v x 1.732 = 54.06 primary **line** amps   54.06a/1.732 = 31.2 primary **phase** amps

1. **(c)**

2. **(b) 0.96**   12,442w/208v x 36a x 1.732 = .959 or .96

3. **(a) 12**

PHASE A	PHASE B	PHASE C
6.6 amps	6.6 amps	6.6 amps
6.6 amps	6.6 amps	6.6 amps
13.2 amps	13.2 amps	13.2 amps

4. **(a) phase amps x 1.732**   III. is **not** correct because of **k**va, if it was va it would also be correct

5. **(c) 32 kw**

PHASE A	PHASE B	PHASE C
32 kw	~~32 kw~~	~~32kw~~

6. **(a) equal to the secondary phase current**

THREE-PHASE EXAM 10 ANSWERS

1. **(c) largest**

2. **(b) 60 amp** 50,000va/480v x 1.732 = 60 amperes

3. **(a) 50 amps** 6000w/120v = 50 amperes

PHASE A	PHASE B	PHASE C
2000w	2000w	2000w
2000w	2000w	2000w
2000w	2000w	2000w
6000w	~~6000w~~	~~6000w~~

4. **(b) 36 amps** 10,000va/480v = 20.8 phase amps x 1.732 = 36 line amps

5. **(c) 36 amps** I =E/R = 208v/10Ω = 20.8 phase amps x 1.732 = 36 line amps

6. **(d) 144 amp** Phase B current = 10,000va/240v = 41.7a Phase C current = 30,000va/240v =125a
L3 current = 41.7a + 125a = 166.7a/2 = 83.35a x 1.732 = 144 amperes

By Tom Henry

12 VIDEOS FOR ELECTRICAL EXAM PREPARATION!

| 12 VIDEO TAPES EXPLAINING EXAM CALCULATIONS- ALL TAPES ARE APPROXIMATELY 75-90 MINUTES |

THE EXAM
Video #301

OHMS LAW - THEORY
Video #302

VOLTAGE DROP - RESISTANCE
Video #303

AMPACITY CORRECTION FACTORS
Video #304

MOTORS
Video #305

COOKING EQUIPMENT DEMAND FACTORS
Video #306

DWELLING-RESIDENTIAL SERVICE SIZING
Video #307

BOX and CONDUIT SIZING
Video #308

SINGLE-PHASE TRANSFORMERS
Video #309

THREE-PHASE TRANSFORMERS
Video #310

COMMERCIAL-MULTIFAMILY SERVICE SIZING
Video #311

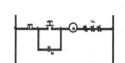

MOTOR CONTROL - SWITCH CONNECTIONS
Video #312

EACH VIDEO $29.95
JOURNEYMAN SERIES $250.00
ITEM # 313 (Includes tapes #301 through #309 total of 9 videos- SAVE $$$$)

MASTER SERIES $350.00
ITEM # 314 (All 12 video tapes- SAVE $$$$)

AUDIOS
By Tom Henry

3 AUDIO TAPES FOR ELECTRICAL EXAM PREPARATION!

The troublesome CLOSED BOOK is now made easy as Tom Henry goes over each answer in DETAIL on audio tape. 3 CLOSED BOOK exams with **150** questions selected by Tom Henry to help the exam applicant learn this part of the electrical examination. Each CLOSED BOOK EXAM contains 50 questions. A ONE HOUR tape narrated by Tom Henry gives in full detail the answer to each question on each exam. A total of 3 hours of audio! You will learn how to memorize and store items in your mind. These exams can be worked over and over again. **The 3 audio tapes can be listened to as you drive in your truck or as you sit at home.** The student will memorize burial depths, definitions, service clearance heights, etc. With **ITEM #192** you will receive a workbook containing 3 closed book exams plus 3 audio tapes by *Tom Henry*. Price $29.00

 Call 1-800-642-2633 Today!

Tom Henry's TWO BRAND NEW BOOKS!

INTRODUCTORY OFFER COMBO $30

 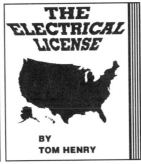

ITEM #276 - The Electrical Plan Reading Workbook   **$18.00**

An excellent workbook for training programs as it starts with symbols and abbreviations and takes the student a walk through the floor plan, panelboard schedule, load balancing schedule, neutral balancing, riser diagrams, one-line drawings, short-circuit calculations, conductor withstand ratings, load balancing delta, load balancing wye, sizing panelboards, slab drawings, detail drawings, specifications, material designations, and much more. The workbook includes 14 exams with answers to test what has been learned. These pages are perforated for removal by the instructor. *Knowledge of plan reading is now required on most electrical exams.*

ITEM #277 - The Electrical License   **$15.00**

This book contains information from all 50 states and several large cities on the requirements for taking the electrical exam. The book lists who writes the exam, the exam format, the exam dates, reciprocity with other states, etc. The book also states if the Business and Law exam is required and if Continuing Education is required for recertification of your license. Lists the address and phone numbers for the contact person in each state.

ITEM #278 - SPECIAL 2-BOOK COMBO   **$30.00**

Make your library complete! Purchase both of Tom Henry's brand new books "Electrical Plan Reading Workbook" and "The Electrical License" for only $30 and save $5 over individual prices.

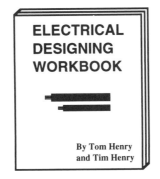

"The Book Every Contractor Should Have On His Desk"

THE BOOK FOR THE FINAL YEAR APPRENTICE

LEARN THE CORRECT APPLICATION OF THHN

ITEM #283 - ATTENTION TRAINING DIRECTORS and SCHOOLS! The *"ELECTRICAL DESIGNING WORKBOOK"* is the book Tom Henry has been working on for years. Along with his son Tim, every effort has been made to design the circuits the electrician installs each day. This book will show you how to select the proper size conductors using all the variables such as continuous loading, voltage drop, ambient temperature, Note 8, cost factors, proper use of THHN conductors connected to 60° or 75° terminations, conductor withstand ratings, neutral sizing, load balancing, panel board sizing, harmonics, sizing bonding jumpers, gutter sizing, and much more. This book is designed for electrical apprenticeship training programs and schools. The workbook contains exams on each chapter. Hundreds of on the job type questions to test the student's knowledge. **Electrical School programs qualify for discount prices.** *THE BOOK!* Price **$35.00**

 CALL TODAY 1-800-642-2633

...Start your complete electrical library with *Tom Henry* Publications!!!

Book #101 - An ideal book for an electrician needing a refresher on Ohm's Law. Explains AC and DC in layman terms with easier to understand formulas, sketches of circuits, series & parallel circuits, Exact K for voltage drop calculations, function of neutral, review on math for the electrician! Contains over 100 Questions & Answers.

Book #102 - An excellent study-aid for the helper, apprentice, or electrician to prepare for the Journeyman license exam. The book contains **10** closed book exams and **12** open book exams. Over **1100** actual exam questions with answers and Code references. An excellent book to study the Code!!

Book #103 - A book designed to advance the electrician in the Code book from the Journeyman level. Contains **8** closed book exams and **10** open book exams. Over **1100** actual exam questions with answers and Code references. An excellent study-aid, takes you cover to cover in the Code including exceptions and Fine Print Notes!!

Book #104- Finally a book written by an electrician in an easy to study format to prepare the everyday electrician in this difficult area of the exam. Single-phase, three-phase, delta-wye, load balancing, neutral calculations, open-delta, high-leg delta, etc. Over 100 calculations with answers!!

Book #105 - Tom Henry's TOP SELLER!! Everything on calculations- **8** chapters - Cooking equipment demands, single-phase ranges on a 3 phase system, ampacity, box-conduit fill, motor circuits, service sizing, feeder sizing, cable tray calculations, and mobile homes, etc. A must!

Book #110 - Tom Henry's favorite reference book. A complete reference book for the electrician that gives the definitions of the language used in the construction field. Also contains formulas used for the exam and in the field, diagrams showing motor, transformer, and switch connections, etc...

Book #115 - The Electrical Alarm Contractor Exam Workbook designed to prepare you for the burglar and fire alarm exam. Hundreds of exam questions with answers. OSHA, UL, NFPA, Life Safety Code, Business Law, loop circuits, etc. !!!!!!

Book #171 - Tom Henry's quiz book! 60 quizzes on tool identification, wiring methods, blueprint symbols, meter reading, circuit testing, controls, proper installation, etc. This book was written to help prepare an electrician for the mechanical, comprehension and aptitude testing of the exam.

Book #197 - The Grounding Workbook designed for training programs to take the mystery out of grounding. Tom Henry's favorite book! Every person in the electrical industry should work the 27 exams this book contains.
Book #199 - INSTRUCTORS GUIDE

Book #198 - The Pictorial Workbook of the Code - Volume One. The Code book in pictures. Volume One starts at the beginning: Articles 90, 100, 110, 200, 210 & 215. A must book for every electrical training program! Now you can "learn" the Code!
Book #201 - INSTRUCTORS GUIDE

Book #108 - KEY WORD INDEX. Every word in the 1993 Code book put into an index with page numbers. Now you can find it in the Code in **seconds**! This is the book the electricians are raving about. Don't be without one!

Book #236 - The Pictorial Workbook of the Code - Volume Two. Volume Two starts at Calculations Article 220. This Volume also includes Outdoor Circuits Article 225, Services Article 230 and Overcurrent Protection Article 240. Volume Two has 25 exams and four final Exams.
Book #237 - INSTRUCTORS GUIDE.

Item #111 - Tom Henry's 1993 Code Tabs. Have all the **KEY** Code References at your finger tips! A special row of service calculation tabs for both residential and commercial. 6 motor calculation tabs to size the wire, heaters, breakers, feeders, etc. to motors. Contains a total of 68 tabs.
Will fit all types of the N.E.C.

Book #107- Control Circuits. A "most requested" book to make control circuits easy to understand for the electrician that is unfamiliar with controls. Circuits are drawn with pictures even showing the flow of voltage throughout the schematic. Order yours today!

Book #106 - "Above The Ceiling". By Popular Request Tom Henry has made his book on humor. Humor that has been collected from over 15,000 Electricians over the past 36 years! Hundreds of "one liners" with graphics.

Book #109 - How To Pass The Electrical Exam. Why take an electrical exam when you are not prepared and fail. Failing hurts, plus it's expensive re-taking exams. This book explains how to read, how to memorize and what to memorize. It breaks down each part of an exam and shows you exactly how to prepare for it. Now you can test yourself to see if you are prepared to take an exam.

Item #116- Formula Insert Pages. Tom Henry's 12 pages of calculation formulas and formats. Formulas for exact K, voltage drop, efficiency, ohms law, kva, transformers, ambient corrections, motor calculation steps, etc.... Pages are predrilled to fit the Looseleaf Code book. Excellent guide to have in your Code book.

Book #212 - "REMINDERS for the Electrician" Book which contains the hard to remember load calculation formats, bus bar formulas, neutral balancing formulas, reversing connections of motors (split phase, capacitor, wound rotor, synchronous, etc.) transformer connections, dwelling formats, cooking equipment formats, switch connections, motor control connections, etc., etc.

Book #276 - The Electrical Plan Reading Workbook - An excellent workbook for training programs as it starts with symbols and abbreviations and takes the student a walk through the floor plan, panelboard schedule, load balancing schedule, neutral balancing, riser diagrams, one-line drawings, short-circuit calculations, conductor withstand ratings, load balancing delta, load balancing wye, sizing panelboards, slab drawings, detail drawings, specifications, material designations, and much more. The workbook includes 14 exams with answers to test what has been learned. These pages are perforated for removal by the instructor.

Book #277 - The Electrical License Book - This book contains information from all 50 states and several large cities on the requirements for taking the electrical exam. The book lists who writes the exam, the exam format, the exam dates, reciprocity with other states, etc. The book also states if the Business and Law exam is required and if Continuing Education is required for recertification of your license. Lists the address and phone numbers for the contact person in each state.

Tom Henrys's Code Electrical Classes Inc. & Bookstore - 6832 Hanging Moss Road - Orlando, FL 32807 - 1-800-642-2633